AF761936

Green Adhesives

Scrivener Publishing
100 Cummings Center, Suite 541J
Beverly, MA 01915-6106

Publishers at Scrivener
Martin Scrivener (martin@scrivenerpublishing.com)
Phillip Carmical (pcarmical@scrivenerpublishing.com)

Green Adhesives

Preparation, Properties and Applications

Edited by
Inamuddin, Rajender Boddula, Mohd Imran Ahamed and Abdullah M. Asiri

Scrivener Publishing

WILEY

This edition first published 2020 by John Wiley & Sons, Inc., 111 River Street, Hoboken, NJ 07030, USA and Scrivener Publishing LLC, 100 Cummings Center, Suite 541J, Beverly, MA 01915, USA
© 2020 Scrivener Publishing LLC
For more information about Scrivener publications please visit www.scrivenerpublishing.com.

All rights reserved. No part of this publication may be reproduced, stored in a retrieval system, or transmitted, in any form or by any means, electronic, mechanical, photocopying, recording, or otherwise, except as permitted by law. Advice on how to obtain permission to reuse material from this title is available at http://www.wiley.com/go/permissions.

Wiley Global Headquarters
111 River Street, Hoboken, NJ 07030, USA

For details of our global editorial offices, customer services, and more information about Wiley products visit us at www.wiley.com.

Limit of Liability/Disclaimer of Warranty
While the publisher and authors have used their best efforts in preparing this work, they make no representations or warranties with respect to the accuracy or completeness of the contents of this work and specifically disclaim all warranties, including without limitation any implied warranties of merchantability or fitness for a particular purpose. No warranty may be created or extended by sales representatives, written sales materials, or promotional statements for this work. The fact that an organization, website, or product is referred to in this work as a citation and/or potential source of further information does not mean that the publisher and authors endorse the information or services the organization, website, or product may provide or recommendations it may make. This work is sold with the understanding that the publisher is not engaged in rendering professional services. The advice and strategies contained herein may not be suitable for your situation. You should consult with a specialist where appropriate. Neither the publisher nor authors shall be liable for any loss of profit or any other commercial damages, including but not limited to special, incidental, consequential, or other damages. Further, readers should be aware that websites listed in this work may have changed or disappeared between when this work was written and when it is read.

Library of Congress Cataloging-in-Publication Data

ISBN 978-1-119-65504-6

Cover image: Pixabay.Com
Cover design by Russell Richardson

Set in size of 11pt and Minion Pro by Manila Typesetting Company, Makati, Philippines

10 9 8 7 6 5 4 3 2 1

Contents

Preface

Synthetic adhesives are known for their toxic impact on the natural environment. In the emerging scientific world, many challenges are arising in the industrial sectors as a consequence of trying to meet the demands posed by upgraded technologies. The development of green adhesives based on renewable resources is the necessity of this age. Their importance is growing gradually as the commercial market is motivated to emphasize the benefits of green adhesives. Green adhesives, particularly those based on polysaccharides, have a number of good features, like biodegradability, biocompatibility, bio-inertness, antimicrobial activity, nontoxicity, and low cost, that keep pace with synthetic adhesives.

This edition of *Green Adhesives: Preparation, Properties and Applications* deals with the fabrication methods, characterization, and applications of green adhesives. It also includes the collective properties of waterborne, bio, and wound-healing green adhesives. Exclusive attention is devoted to discussing the applications of green adhesives in biomedical coatings, food, and industrial applications. This book will be useful for beginners and experts from undergraduate students to industrial engineers working in the field of polymer chemistry, materials science, and engineering. Based on thematic topics, this edition contains the following ten chapters:

Chapter 1 provides an overview of current research progress, major challenges, and future perspectives of the developed anti-adhesive coatings, including hydrophilic polymers, zwitterionic polymers, superhydrophobic polymers, slippery liquid-infused porous surfaces (SLIPS), and protein or glycoprotein-based coatings. Three bifunctional coatings with anti-adhesive and antibacterial activities are also discussed.

Chapter 2 focuses on adhesives that are synthesized with renewable lignin. The structure and reactivity of various industrial lignins are introduced in detail. The methods for modifying technical lignins to improve their reactivity for the synthesis of bio-based adhesives and the performances of resultant adhesives are discussed.

Chapter 3 describes the predominant role of green adhesives in various industrial applications. It further enumerates various categories of green adhesives by pointing out its vital material characterization concerning robotics, microelectronics, and space applications. Moreover, its contribution to powder coating, leather tanning, fiberboard, and dental fields are also discussed.

Chapter 4 highlights some green adhesives produced from polysaccharides or biopolymers which are widely available and play an important role in biomedical systems for various applications such as tissue engineering, drug and gene delivery, and wound healing. Several physicochemical characterizations of biopolymers and the main advantages and feasibility of some procedures found in the literature also are discussed.

Chapter 5 describes the procedures for modifying polymer systems into waterborne ones by the incorporation of hydrophilic segments by various means. It also demonstrates the structure and properties of waterborne epoxy and polyurethanes that could be used in adhesive applications.

Chapter 6 discusses the preparation of furfuryl alcohol-based adhesives. The methods of preparing polyfurfuryl alcohol-based sealants without blistering based on temperature, amount of the catalyst, as well as thickness of the end products are discussed in detail and validated by mathematical model.

Chapter 7 deals with the overview of bioadhesives, their history, and classification. It also elaborates the mechanism of bioadhesion along with the testing methods and application of bioadhesives. Focus is placed on bioadhesion mechanisms like sealants, adhesives, hemostatic proxies, etc. This chapter may open doors for researchers working on metals, alloys, and composite materials by introducing a better way of avoiding the failure of the elements using bioadhesives.

Chapter 8 details the development of polysaccharide adhesives based on renewable resources.

Chapter 9 describes the role of adhesives in the healing of wounds that appear after surgical incisions, endoscopic processes, dental procedures, ocular surgeries of both cornea and retina, cataract, pancreatectomy, gastric incisions, inguinal hernia repair, variceal bleeding, knee arthroplasty, cardiac surgeries, diabetic foot ulcers, accidental lacerations, and burns.

Chapter 10 discusses synthetic or renewable adhesive systems that can replace formaldehyde-based adhesives. In this chapter, recent research on wood flooring adhesives is included.

Chapter 11 focuses on the basic characteristics, classification, production, benefits, properties, application of synthetic adhesives and resins for bonding glass, metals, ceramics, wood, rubber, foil, paper, leather, cloth, plastic films, and some structural materials.

Editors
Inamuddin
Rajender Boddula
Mohd Imran Ahmed
Abdullah M. Asiri

1

Anti-Adhesive Coatings: A Technique for Prevention of Bacterial Surface Fouling

Xiaohong Sun[1]*, Songyuan Zhang[1], Hui Li[2] and Nandika Bandara[3]

[1]College of Food and Biological Engineering, Qiqihar University, Qiqihar, China
[2]Institute of Quality Standard & Testing Technology for Agro-Products, Chinese Academy of Agricultural Sciences, Beijing, China
[3]Department of Plant, Food & Environmental Sciences, Faculty of Agriculture, Dalhousie University, Truro, Canada

Abstract

Bacterial surface fouling (biofouling) causes many severe problems in a range of applications, such as biomedical field and food industry, which can result in enormous economic losses, pose threats to human health, and even lead to death. Protein adsorption on surfaces often serves as a conditioning layer for initial bacterial adhesion that is considered as the first and prerequisite step for biofouling. Therefore, anti-adhesive coatings have been designed as an effective approach to prevent biofouling. This chapter provides an overview of the current research progresses, major challenges and future perspectives on the developed and generated anti-adhesive coatings, including hydrophilic polymers, zwitterionic polymers, super-hydrophobic polymers, slippery liquid infused porous surfaces (SLIPS), and protein or glycoprotein-based coatings. Also, the working principles and efficacies of three bifunctional coatings with anti-adhesive and antibacterial activities are further discussed. Overall, anti-adhesive coatings show great potential to be used to minimize the negative impacts of biofouling on different practical applications, such as biosensors, biomedical implants, as well as pipes, equipment, and cooling systems in food industry. However, it is still demanding to develop safe, simple, economical, easily available, durable, and effective anti-adhesive coatings in future.

Keywords: Anti-adhesive coating, bacterial adhesion, protein adsorption, biofouling, antibacterial property, bifunctional coating, biofilm

**Corresponding author*: xs7@ualberta.ca; sunxiaohong55@hotmail.com

Inamuddin, Rajender Boddula, Mohd Imran Ahamed and Abdullah M. Asiri (eds.) Green Adhesives: Preparation, Properties and Applications, (1–24) © 2020 Scrivener Publishing LLC

1.1 Bacterial Surface Fouling (Biofouling)

Biofouling is an undesired accumulation of biomacromolecules (e.g., proteins) or organisms (e.g., bacteria) on material surfaces [1]. More than 99% of bacteria on earth are surface-attached cells and can form biofouling [2, 3]. The process of bacterial surface fouling mainly includes conditioning film development, initial adhesion, and subsequent biofilm formation. A layer of organic molecules (e.g., proteins and polysaccharides) adsorption on surfaces serves as a conditioning film for bacterial adhesion [1]. The initial bacterial adhesion to the surfaces is partially reversible since it can be more readily disrupted compared to the strongly adhering biofilms [4, 5]. Biofilms are defined as the communities of microorganisms embedded in extracellular polymeric substances (EPS), which can increase the resistance to host immune responses, mechanical removal, and antibiotic treatments compared to planktonic cells [6, 7].

Initial bacteria-surface attachment and adhesion are considered as the prerequisite step for biofilm formation, which involved several mechanisms, including hydrophobic and electrostatic interactions [1, 3]. Environmental conditions, such as pH, temperature, calcium ions, competing organisms and nutrition levels, and bacterial sizes, shapes, and surface structures such as pili, flagella, fimbriae, and adhesins, are demonstrated to significantly impact on this initial adhesion [8, 9]. Besides, the material surface properties play a vital role in this initial adhesion as well, including surface chemistry (e.g., functional groups and coatings), charge, energy (mainly related to the surface hydrophobicity), structure (size, spacing, aspect ratio, and roughness), stiffness, and topography. Compared to other factors, the influences of surface stiffness and topography on the bacterial adhesion have not been well studied [3, 4, 8, 10]. Overall, the process of bacterial adhesion is complicated, which governed by all of the factors discussed above.

1.2 Negative Effects of Biofouling by Bacteria on Practical Applications

Bacterial adhesion and formation of biofilm on medical implants (such as prosthetic joints or catheters) cause device-related infections, which has long been a significant problem in the biomedical field, both during and after implant surgery [11, 12]. Implant-related infections may result in a high occurrence rate of revision surgeries, and even fatality, placing a greater financial burden on the healthcare system of many countries

[12–14]. Taking the United States as an example, the direct medical costs associated with such infections was estimated more than $3 billion annually about one decade ago, which is expected to increase nowadays [15].

In the food industry, biofouling formation, especially biofilm, in equipment, pipes, and cooling systems causes enormous economic losses through increasing maintenance costs and reducing equipment operational efficiencies [16, 17]. Biofilm can also be a source of microbial contamination, causing food-borne infectious diseases and posing a threat to human health. Food-borne diseases are still a global public health concern nowadays, and new ones continue to emerge [18, 19]. Also, microbial contamination negatively impacts on food storage, which results in undesirable changes to the sensory and quality of food [20]. Currently, it is estimated that 25% of the total food produced annually wasted due to microbial contamination [20].

1.3 Anti-Adhesive Coatings for Preventing Bacterial Surface Fouling

As discussed above, the initial bacteria adhesion to the surfaces is easier to be interfered compared to the strongly adhering biofilms. Therefore, many anti-fouling strategies were designed to prevent the initial bacterial adhesion and biofilm formation on surfaces instead of removal of the mature biofilms, for example, fabrication and identification of anti-adhesive coatings. Different types of polymers have been applied to prepare anti-adhesive coatings, such as hydrophilic polymers, zwitterionic polymers, super-hydrophobic polymers, slippery liquid infused porous surfaces (SLIPS), and protein or glycoprotein-based coatings. This chapter will provide an insight into the current research progresses, major drawbacks, and future directions in the field of anti-adhesive coatings.

1.3.1 Hydrophilic Polymers

It is well accepted that the application of hydrophilic polymers as coatings on surfaces could resist adhesion of proteins and bacteria due to the formation of a hydration layer in an aqueous environment as a physical barrier [21]. Schematic diagram of anti-adhesive coatings based on hydrophilic polymers was given in Figure 1.1a. Two widely studied hydrophilic polymers will be introduced below, including polyethylene glycol (PEG) and poly(2-methyl-2-oxazoline) (PMOXA).

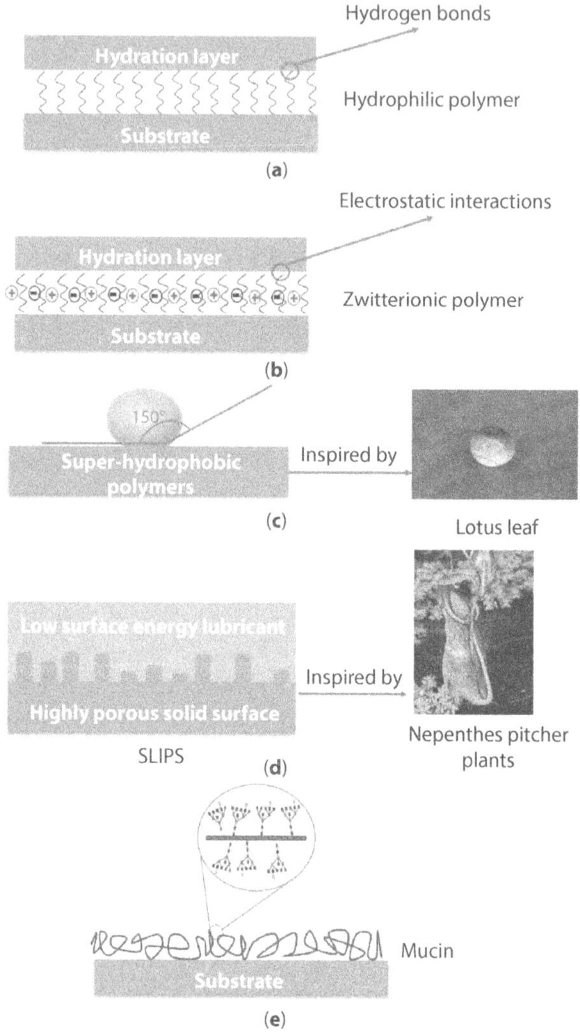

Figure 1.1 Schematic diagrams of five types of anti-adhesive coatings. (a) Hydrophilic polymers as coatings on the substrate resist biofouling due to the formation of a hydration layer in aqueous environment as a physical barrier and the surface hydration layer is formed by weak hydrogen bonds; (b) Forming a hydration layer on the substrate by strong electrostatic interactions is mainly responsible for the anti-biofouling property of zwitterionic coatings; (c) The major characteristic of super-hydrophobic surface is with a water contact angle of over 150°, which was inspired by the Lotus leaf in nature [72]. Super-hydrophobicity can reduce the adhesive force between bacterial and the surface; (d) Slippery liquid infused porous surfaces (SLIPS) were inspired by Nepenthes pitcher plants, which are commonly composed of a highly porous solid surface and a low surface energy lubricant; (e) Glycan moieties and critical surface density of mucins are indispensable to achieve the anti-adhesive activity.

Polyethylene glycol (PEG)-based coatings

PEG is one of the most well-documented anti-fouling polymers and has been recognized as the gold standard for anti-fouling coatings [22]. Coatings-based on PEG has been used in a variety of applications, such as biosensors and tissue engineering [23, 24]. The unique physical and biochemical properties of PEG are critical for its anti-fouling applications, like non-toxicity, non-immunogenesis, non-antigenicity, and good biocompatibility [25, 26].

PEG showed good anti-fouling activity against platelet adhesion and protein adsorption. For instance, Zhang *et al.* [27] modified the gold surface by grafting of PEG polymer and further characterize this modified surface by the techniques of X-ray spectroscopy, atomic force microscopy (AFM), and water contact angle. It was demonstrated that at a surface [C-O]/[Au] ratio above 50, PEG modified-gold surface could completely prevent protein adsorption and platelet adhesion, and such surface can be stable in water and protein solutions for more than 24 h. Likewise, PEG-modified silicon surfaces exhibited potent activity to reduce protein adsorption, including bovine serum albumin and fibrinogen [28]. Besides, it was proved that polyethylene oxide (PEO) which is the same polymer as PEG since they have the same repeating unit ($-CH_2CH_2-O-$) could efficiently reduce platelet adhesion and fibrinogen adsorption as well [29–31]. It was postulated that steric barrier, chain length, surface density, refractive index, and hydrophilic nature of polymers are responsible for protein resistance [32–34].

It was hypothesized that, since a layer of adsorbed protein can act as a conditioning film for bacterial adhesion, surfaces with anti-fouling property should prevent bacterial adhesion [35]. Therefore, much research efforts have been devoted to examining this hypothesis and demonstrated that PEG-based surfaces could resist bacterial adhesion [36–38]. It was reported that a commercial multi-component cross-linked PEG-based coating (OptiChems) significantly inhibited adhesion of all clinically isolated strains tested in the study, including *Staphylococcus epidermidis* (*S. epidermidis*) 3399, *S. epidermidis* HBH 276, *S. salivarius* GB 24/9, *Pseudomonas aeruginosa* (*P. aeruginosa*) #3, except *Escherichia coli* O_2K_2, and the inhibitory rates were all more than 80%. The authors concluded that exposure of OptiChems coating into protein-rich fluids, such as saliva and blood plasma, could dramatically decrease its anti-adhesive activity although the coating integrity was not interfered by proteins [36]. The anti-adhesive property of PEG-based coatings strongly relied on the PEG immobilized strategy; namely, covalent immobilization was better than physical adsorption [37].

Further, the forces of interaction between bacteria and PEG-coated surfaces were measured by AFM to indicate the possible anti-fouling mechanism. The data suggested that PEG chains not only interfered the long-range attractive forces between bacteria and substrates but also exhibited large steric repulsion forces [39]. Also, the steric repulsion and dynamic motion of hydrated polymer chains as well as chain length are important factors in preventing bacterial adhesion [22, 40, 41].

However, surfaces based on PEG cannot entirely prevent bacterial adhesion/colonization. A recent study demonstrated that the polymer brush coating of poly(L-lysine)-graft-PEG (PLL-g-PEG) on titanium did not resist *S. epidermidis* colonization and biofilm formation. Because the biofilm formed by *S. epidermidis* was rich in polysaccharides and extracellular DNA, which could interact with PEG brushes and completely remove the polymer from titanium surface. This drawback could be overcome by increasing the polymer brush density [42]. Furthermore, PEG has several obvious limitations: susceptible to oxidation in complex media, not suitable for long-term applications, not biodegradable, bioaccumulation in the lysosomes of healthy cells, provoking an immune response in some circumstances [22, 43–45].

Many researchers have attempted to improve the characteristics and anti-fouling activity of PEG polymer. PEG-grated polyurethane surfaces, which were fabricated by combining chemical and physical approaches, showed improved biocompatibility and increased activity to inhibit both platelet adhesion and bacterial adhesion/biofilm formation [46]. A layer-by-layer assembling technique has been used to generate the PEG containing multilayer films which showed potent anti-adhesive activity. For example, PEG grafted poly (L-lysin) (PLL) or poly (L-glutamic acid) (PGA) possessed potent activity to notably suppressing both protein adsorption and bacterial adhesion [47].

Poly(2-methyl-2-oxazoline) (PMOXA)

PMOXA is highly soluble in water and has attracted increasing interest as an alternative to PEG. A direct comparison between PEG and PMOXA with respect to protein adsorption, bacterial adhesion, and oxidative stability was well-reviewed. Overall, PMOXA coatings showed equally excellent anti-fouling activity as that of PEG and PMOXA is more stable in oxidative environments [48].

Specifically, PEG and PMOXA-based polymers were equally effective in inhibiting protein adsorption from whole human serum according to a quantitative comparison [49]. The side-chain grafting densities of polymers played an important role in the adsorbed masses of polymer and protein on surfaces. The optimum side-chain grafting density relies on the side-chain molecular weight. In general, the higher the molecular weight, the lower is the optimum grafting density [50]. As such, the optimum grafting density of PMOXA-based polymer is approximately half the density of PEG-based polymer since the molecular weight of monomer unit (N-acetylethyleneimine) of PMOXA is twice as compared with ethylene glycol [49].

Like for the prevention of protein adsorption, both PEG and PMOXA-based coatings had an equal potency in inhibition of bacterial adhesion [49]. To measure the anti-adhesion, polymer coatings were exposed in a bacterial suspension of *Escherichia coli* (*E. coli*) K12 at a concentration of 10^9 cfu/mL in physiological buffer at 37°C. The results showed that both PEG and PMOXA-based coatings could reduce more than 3 log units of bacterial adhesion [51].

Nevertheless, PMOXA revealed a significantly higher oxidative stability over PEG. It was demonstrated that PEG monolayers were lost more than 50% of their initial thickness when exposed to the oxidative environment, while the initial thickness of PMOXA monolayers was only degraded less than 20%. Furthermore, the degraded PEG coatings largely lost the capacity for prevention of protein adsorption, while PMOXA remained a good activity to resist protein adsorption after oxidation [49].

1.3.2 Zwitterionic Polymers

Zwitterionic polymers are a class of materials that show overall electrical neutrality since they have an equal number of negatively and positively charged groups within a single molecule [29]. Zwitterions were inspired by the non-thrombogenic property of zwitterionic lipid phosphatidylcholines on the outer membrane of erythrocyte cells [52]. The general preparation method of zwitterionic anti-biofouling polymers is incorporating cationic and anionic compounds into the backbone of polymeric materials. The most well-investigated zwitterionic polymers are poly(carboxybetaine) (pCB) and poly(sulfobetaine) (pSB) grafted to methacrylate or acrylamide backbones, which are stable and easy to deal with [21, 53]. Similar to hydrophilic polymers, forming a hydration layer on the surface plays a major role in the anti-biofouling property of zwitterionic coatings. However, in contrast to hydrophilic surfaces whose surface hydration layer is formed by weak hydrogen bonds, charged zwitterionic polymers maintain the hydration layer through strong electrostatic interactions, as shown in Figure 1.1b. It is believed that the more tightly possessed the hydration layer, the more potent the material is at resisting a disruption and biofouling adhesion [21, 29].

The features of overall charge neutrality, high hydrophilicity, strong dipole pairs, and anti-polyelectrolyte effect endow zwitterionic polymers with promising applications for fighting against biofouling problems [54]. Phospholipid polymers, poly (2-methacryloyloxyethyl phosphorylcholine) (MPC)-co-n-butyl methacrylate (BMA), were the first zwitterionic material to be proved for the anti-biofouling property, which was reported by Ishihara *et al.* in 1991 [55]. However, this was only a preliminary research and the poly(MPC-co-BMA) polymer only resisted protein adorption from single protein solutions. Since then, more zwitterionic polymers have been developed and demonstrated to possess potent anti-biofouling capacity of suppressing nonspecific protein adsorption from complex media. For example, gold surfaces coated by zwitterionic poly[2-(methacryloyloxy)ethyl phosphorylcholine] have revealed an excellent protein-resistant property which was comparable to that of PEG-based coatings [56]. Moreover, Yang *et al.* [57] firstly demonstrated that poly(carboxybetaine acrylamide) (polyCBAA)-grafted surfaces with a film thickness of ~21 nm that were prepared via atom-transfer radical polymerization (ATRP) could achieve "zero" protein adsorption from complex undiluted human blood serum and plasma.

The resistance to bacterial adhesion and biofilm formation of zwitterionic polymers were further investigated. For instance, as compared to the glass reference, long-chain zwitterionic poly(sulfobetaine methacrylate) (pSBMA) surfaces not only significantly reduced short-term (3 h) adhesion of *S. epidermidis* and *P. aeruginosa*, also notably impeded their biofilm formation [58]. The effect of zwitterionic poly(carboxybetaine methacrylate) (pCBMA)-grafted surfaces on long-term bacterial biofilm formation was further investigated. Bare glass coverslips were selected as a reference. Results showed that pCBMA coatings suppressed long-term biofilm formation of *P. aeruginosa* up to 240 h by 95% at 25°C and 64 h by 93% at 37°C. For strain *P. pudida*, pCBMA surfaces maintained reducing the biofilm formation up to 192 h by 95% at 30°C [59]. Similarly, zwitterionic micellar coatings and zwitterionic polymers with a biomimetic adhesive group showed strong anti-adhesive activity against *S. aureus* and *P. aeruginosa*, respectively [60, 61].

In addition to those mentioned above, zwitterionic polymer-grafted surfaces and zwitterionic hydrogels also exhibited potent anti-biofouling activity. The anti-adhesive activity of zwitterionic pCBMA hydrogel has been most widely explored. pCBMA hydrogels retarded cell adhesion by 90% compared to control hydrogels (2-Hydroxyethyl methacrylate) and the pCBMA hydrogels can be stable in 100% human blood serum up to 12 days [62]. Meanwhile, antimicrobial agents, such as salicylic acid (SA), were conjugated to zwitterionic hydrogels to achieve dual anti-biofouling and antibacterial activities because SA can be released from hydrogel network to inhibit bacterial growth on the surface and surrounding media [63]. Zwitterionic hydrogels are believed to have promising applications in biosensors, contact lenses, and biomedical implants [54].

Although zwitterionic polymers have been regarded as an attractive alternative to PEG-based coatings, their hydration layers will be disrupted over time, and surface biofouling can reoccur. To this end, more research efforts are needed to develop more durable zwitterionic coatings in the future.

1.3.3 Super-Hydrophobic Polymers

Inspired by the Lotus leaf in nature, the super-hydrophobic surface has been designed and fabricated in the last two decades. The major characteristic of the super-hydrophobic surface is with a water contact angle (WCA) of over 150°, as illustrated in Figure 1.1c [64]. Super-hydrophobic surfaces are conventionally generated mainly in two approaches. One is to modify a hydrophobic surface (WCA > 90°) with a rough structure, and the other is to establish a rough surface by materials with low surface free energy. Overall, to prepare super-hydrophobic surfaces, coating with low-surface-energy materials such as fluoroalkylsilane is commonly applied [64, 65].

It has been well accepted that generally hydrophobic surfaces are incapable of preventing protein adsorption; on the contrary, proteins tend to adsorb on hydrophobic surfaces [66]. However, the super-hydrophobic surface is an exception which has been demonstrated to have potent activity to low protein adsorption and remove proteins [67, 68]. Since protein layer is a conditional film for bacterial adhesion, it was hypothesized that super-hydrophobic surfaces could prevent bacterial adhesion. Many researchers have made great efforts to test this hypothesis. Tang *et al.* [69] investigated the anti-adhesive activities of a TiO_2-based super-hydrophobic surface (with a WCA of 156°) and a hydrophobic surface (with a WCA of 133°) against *Staphylococcus aureus* (*S. aureus*). The results showed that although the super-hydrophobic surface did not completely inhibit *S. aureus* adhesion, while the number of adherent bacteria on the super-hydrophobic surface were much less than those on the hydrophobic surface. More importantly, adherent bacteria on the super-hydrophobic surface could be easily removed. It was further demonstrated that the super-hydrophobic coating with a WCA of 167° fabricated from fluorinated silica colloids significantly reduced *S. aureus* and *E. coli* adhesion [70]. In addition, super-hydrophobic coating that was prepared by annealing stainless steel plates with carbon nanotubes-polytetrafluoroethylene significantly decreased *E. coli* K12 adhesion (approximately 80%) compared to uncoated surface. The authors indicated that this super-hydrophobic coating has great potential to be used in food industry to limit the risk of cross-contamination [71]. Based on the current studies, the anti-adhesive activity of super-hydrophobic coatings against bacteria is mainly due to the reduced protein adsorption and the entrapped air layer between the bacteria and the surface. The easy removal of bacteria on the super-hydrophobic surface is because super-hydrophobicity can reduce the adhesive force between bacterial and the surface [72].

However, there are some controversial findings. For example, it was reported that both *S. aureus* and *P. aeruginosa* could adhere to the super-hydrophobic poly(L-lactic acid) surface after 24-h incubation and the adherent *P. aeruginosa* could even form biofilms on the surface. And the adherent bacteria were not easily removed by water [73]. The contradictory results may be due to different determination methods for bacterial adhesion and distinct preparation methods for the super-hydrophobic surfaces. Also, super-hydrophobic coatings have many inherent limitations, such as low mechanical stability, and failure under high pressure, temperature, and humidity [74]. Overall, application of super-hydrophobic surfaces to prevent bacterial adhesion is an emerging technique which needs to be studied further in future.

1.3.4 Slippery Liquid Infused Porous Surfaces (SLIPS)

To address the aforementioned problems of super-hydrophobic surfaces, SLIPS were designed in 2011, inspired by Nepenthes pitcher plants [75]. In pitcher plants, there are liquid films that can make insects slide from the rim into the bottom digestive juices due to the oil-repellent property when insects step on it [75]. Hence, SLIPS are commonly composed of a highly porous solid surface and a low surface energy lubricant. The main principle of SLIPS is a steady film of lubricant liquid forms and covers the solid surface due to the special porous topography, thus blocking the contact of other liquids with the underlying surface, which was indicated in Figure 1.1d [76]. The SLIPS are expected to maintain a strong repellent property within a broad range of temperatures, pressures, surface tensions by exploiting the extremely stable liquid film [77].

The anti-biofouling property of SLIPS was firstly confirmed by Epstein *et al.* [77]. In this work, the SLIPS were manufactured by infiltrating a porous fluorinated membrane with a fluorinated fluid, which exhibited potent ability to prevent biofilm attachment of *P. aeruginosa* (99.6%), *S. aureus* (97.2%), and *E. coli* (96%) over a 7-day period under both static and physiologically realistic flow conditions. These results proved that SLIPS showed much stronger anti-biofouling activity than that of PEGylated coatings. The anti-biofouling property of SLIPS is resulting from the exceptional mobility on the slippery interface rather than cytotoxicity. The authors further demonstrated that SLIPS are stable under the conditions of submergence, extreme pH, salinity, and UV exposure. Moreover, SLIPS are low-cost, passive, simple to prepare, and can be fabricated on arbitrary surfaces, which indicated their potential applications such as food and marine industries.

The potential utilization of SLIPS in the food industry was validated by Rungraeng *et al.* [78]. Specifically, SLIPS were fabricated by infusion of a low surface tension liquid lubricant (Krytox) to a superhydrophobic multiwalled carbon nanotube and Teflon composite layer. The prepared SLIPS were coated on the surface of a stainless steel plate for the determination of anti-adhesive activity against *E. coli* K12 and *S. aureus*. The results showed that about 90% reduction in biofilm attachment on the SLIPS compared to the control stainless steel. On the other hand, SLIPS showed an exceptional anti-biofouling activity to inhibit *Chlorella vulgaris*, a typical marine biofouling organism, settlement on surfaces in both static and dynamic conditions [79]. This content will not be elaborated since the main interest of this chapter is in the field of preventing bacterial surface fouling.

Developing generalized methods to prepare SLIPS on the versatile surface is a major challenge. A convenient and inexpensive approach for preparing SLIPS was reported, namely, infusing a lubricating liquid into a nanoparticle-polymer film. The established SLIPS showed various liquids-repellent property, including water, oil, and ethanol [74]. To broaden the potential applications of SLIPS, multi-functional SLIPS concept was introduced by incorporating antimicrobial agents. Antimicrobial agents can be controlled release to the surrounding media, which not only enhanced the anti-biofouling activity of SLIPS, also endowed SLIPS with the ability to kill non-adherent pathogens [80]. However, the research on SLIPS is still at early stage, more well-designed researches on preparation of simple, economical, potent, eco-friendly, and multi-functional SLIPS are highly demanding in future.

1.3.5 Protein and Glycoprotein-Based Coatings

It has been reported that coating of an aqueous fish muscle extract on stainless steel surfaces can significantly inhibit bacterial adhesion [81]. Vejborg *et al.* [82] further identified and characterized the responsible components associated with this anti-adhesive activity using *E. coli* K12 as model bacteria. The results suggested that fish tropomyosins which are fibrous proteins, played an essential role in this activity due to the formation of an anti-adhesive conditioning layer on surfaces. Therefore, fish tropomyosin showed potential to be a non-toxic and relatively inexpensive anti-adhesive coating.

Lubricin is a glycoprotein that is present in articular joints. A recent study demonstrated that lubricin exerted an effective anti-adhesive activity over a wide range of pH, which was comparable to that of self-assembled PEG coatings. Lubricin coatings were most potent on hydrophobic, anionic, and uncharged polar surfaces. Also, lubricin coatings exhibited excellent long-term stability and maintained strong anti-adhesive property at ambient temperatures under both vacuum and atmospheric conditions. The anti-adhesive property of lubricin was possibly resulting from its ordered polymer brush-like layer whereas the grafting density and surface coverage of lubricin coatings were found to be less important.

The main function of the mucous layer that covers epithelial cells is to act as a natural anti-fouling surface to resist undesirable adhesion from hosting tissues [83]. Mucus layer is mainly composed of water and mucin which is a member of heavily glycosylated glycoproteins with a high molecular weight [84]. Therefore, many research efforts have been made to explore the potential of mucin as anti-fouling surface to suppress cell and bacterial adhesion, including bovine submaxillary mucin (BSM) and pig gastric mucin (PGM).

Mammalian mucins, such as BSM and PGM, were reported to adsorb on hydrophobic substrates and form a mucin layer that prevents the surface adhesion to mammalian epithelial cells, fibroblasts, and myoblasts. This cell-repellent effect of mucin coatings could help improve the biocompatibility of biomedical implants. The authors suggested that the glycan moieties and critical surface density of mucins are indispensable to achieve the cell-repulsion [85].

Mucin coatings have exhibited the activity to prevent bacterial adhesion to polymeric material surfaces. In one study, BSM was coated on a surface consisting of a poly(acrylic acid-b-methyl methacrylate) (PAA-b-PMMA) diblock copolymer. BSM coating reduced the adhesion of *S. epidermidis* and *E. coli*; increasing the mucin film thickness led to further decrease of the density of adhering *S. epidermidis* but not *E. coli*. BSM coating had affinity to *C. albicans* but did not resist *C. albicans* adhesion [86]. In another study, BSM was coated to four polymeric materials, including silicone, PMMA, polystyrene, and polyurethane. *S. aureus* and *S. epidermidis* were applied to contaminate the BSM coated surfaces and corresponding bare surfaces. A correlation was established in all BSM-coated polymers: the more BSM adsorption, the fewer bacteria adhesion. The inhibitory percentages of bacterial adherence on BSM-coated PMMA, polystyrene, polyurethane, and silicone were about 97, 90, 92, and 70%, respectively [87]. Coatings of PGM and BSM on polystyrene surfaces suppressed the adhesion of *Streptococcus pneumoniae* and *S. aureus*. The authors further suggested that glycan moieties of mucins are indispensable for bacterial repulsion possibly because glycans are key contributors in the critical thickness and softness of the mucin coatings which are associated with increasing bacterial repulsion [88]. Although mammalian mucin-based coatings showed promising anti-adhesive potential, their massive production is not practical due to high cost and low availability.

Recently, Sun *et al.* [89] determined the protein-resistant property of ovomucin, which is a mucin-type glycoprotein in egg white, to indicate its anti-adhesive potential. Using bovine serum albumin (BSA) as a model protein, ovomucin significantly decreased BSA adsorption on polystyrene surface. To further understand the underlying mechanism, surface forces apparatus, atomic force microscope imaging, zeta potential, and dynamic light scattering were applied in this study. Finally, the authors concluded the protein-resistant property of ovomucin was mainly attributed to the strong steric hindrance and electrostatic repulsion between protein layers. This work has indicated that ovomucin has anti-adhesive potential with broad applications in the fields of biomedical implants and food processing industry, which provided new solutions for developing easily available, inexpensive, and effective anti-adhesive coatings.

1.4 Bifunctional Coatings With Anti-Adhesive and Antibacterial Properties

To date, none of the developed anti-adhesive coatings can completely inhibit the bacterial adhesion. Although anti-adhesive coatings can largely prevent the initial bacterial attachment, it is a passive approach and generally transient since the coatings are not capable of killing bacteria and will be deterioration, making it hard to prevent the ultimate biofouling [90].

This drawback has driven much research in the area of development of bifunctional coatings with both anti-adhesive and antibacterial activities to enhance the anti-biofouling property further. The widely used method to generate the bifunctional coating is to incorporate bactericides to anti-fouling materials. According to different incorporating approaches, three types of bifunctional coatings will be discussed below, and their schematic illustrations were shown in Figures 1.2a–c accordingly.

The first method is to immobilization of bactericidal agents on the surfaces of anti-fouling materials. As discussed above, PEG is deemed as a gold standard of anti-fouling polymers. As such, several antibiotics, including penicillin, ampicillin, and gentamicin, have been tethered on the surfaces of PEG derivatives, such as (poly (tetrafluoroethylene) (ePTFE) that is the NH_2/COOH-terminated PEG, to create bifunctional coatings [91–93]. To achieve the attachment of antibiotics to the ePTFE surfaces, taking ampicillin as an example, the reaction occurs between NH_2 groups of ampicillin and COOH groups of ePTFE [91]. These established coatings showed effective activity to prevent bacterial adhesion and microbial proliferation, including both gram-positive (*S. aureus*, *Bacillus thuringiensis*, and *Enterococcus faecalis*) and gram-negative bacteria (*E. coli*, *Pseudomona putida*, and *Salmonella enterica*) [91–93]. However, antibiotic can easily develop drug-resistant strains which are a substantial global threat to human health. Hence, natural biocides, such as antimicrobial peptides, lysozyme, and chitosan, have been utilized as antibiotic alternatives to prepare the bifunctional coatings, which possessed potent anti-adhesive and antibacterial effects as well [94–96]. In addition, since each grafted ePTFE chain has only one active group at its free end for biocides, PEG-derived polymers with comb-like structures, such as 2-hydroxyethylmethacrylate polymer (PHEMA) and PEG monomethacrylate (PEGMA), were developed for increasing the available binding sites for the antibacterial biomolecules (chitosan or lysozyme) [94, 96].

The second approach is to form a multilayer film using antibacterial materials and anti-adhesive agents with opposite charges by layer-by-layer deposition. This technique is simple, low-cost, and facile [90]. Typically, the anti-adhesive and antibacterial multilayer films are constructed by chitosan (an antibacterial compound with positive charges) and heparin (an anti-adhesive agent with negative charges), which are alternately deposited onto substrates [97, 98]. The formation of the multilayer films via layer-by-layer assembly of heparin and chitosan was validated by infrared spectroscopy (FTIR), atomic force microscopy (AFM), and scanning electron microscopy (SEM) techniques [97]. Results obtained from the anti-adhesive and antibacterial assays using *E. coli* as model bacteria indicated

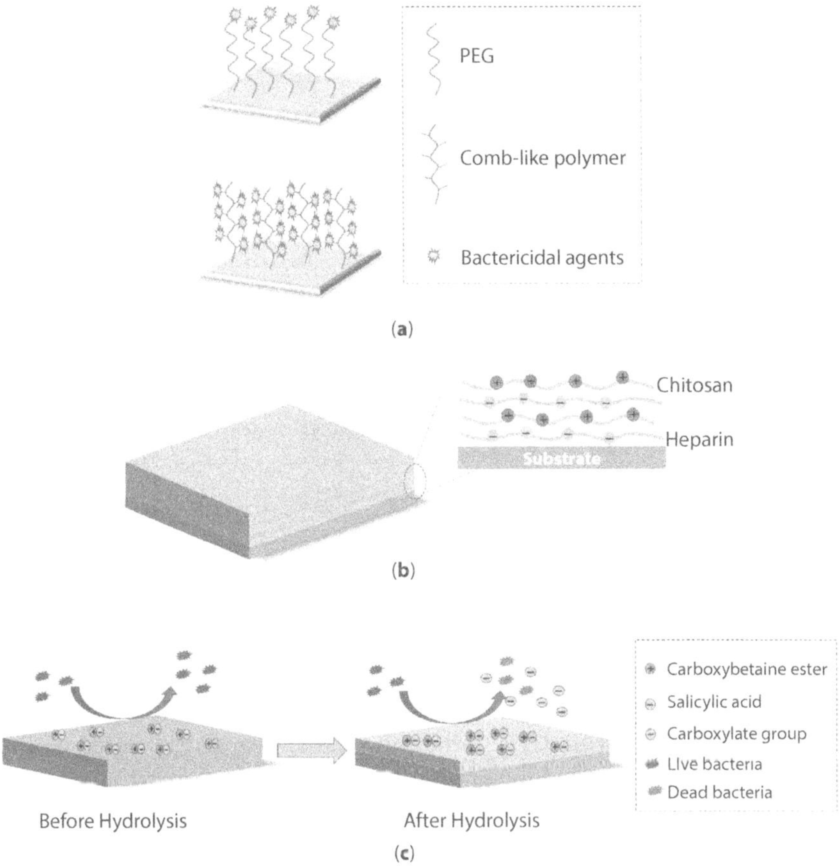

Figure 1.2 Schematic illustrations of bifunctional coatings with anti-adhesive and antibacterial properties, adapted from ref [90]. (a) Bactericidal agents are immobilized on the surfaces of anti-fouling material (PEG), and comb-like polymers increase the available binding sites for the bactericidal agents; (b) The anti-adhesive and antibacterial multilayer films are typically constructed by chitosan (an antibacterial compound with positive charges) and heparin (an anti-adhesive agent with negative charges), which are alternately deposited onto substrates; (c) Salicylic acid, an anionic antibacterial agent, is integrated to a zwitterionic hydrogel (carboxybetaine ester) to keep the overall charge neutrality. When hydrolysis occurs, the released salicylate molecule is replaced by a carboxylate group to maintain the zwitterionic property. The released salicylate kills planktonic bacteria in the surrounding environment.

that the chitosan/heparin multilayer-based coatings significantly suppressed bacterial adhesion and effectively killed the bacteria [97, 98].

Last but not least, antibacterial compounds can also be integrated to zwitterionic hydrogels and subsequently released to prevent bacterial surface fouling and inhibit the planktonic bacterial growth [90]. For instance, salicylic acid, an anionic antibacterial agent, was incorporated into a carboxybetaine ester hydrogel (pCBMA) as an anionic group to balance the cationic quaternary ammonium group in hydrogels to keep the overall charge neutrality [63, 99]. When hydrolysis occurred, the released salicylate molecule was substituted by a carboxylate group to maintain the zwitterionic and anti-biofouling properties. The released salicylate inhibited the growth of planktonic bacteria in the surrounding environment. After the salicylate was completely released, the hydrogel finally changed to a stable zwitterionic form that can further prevent bacterial adhesion. This established pCBMA-salicylate hydrogel inhibited the growth of both gram-negative (*E. coli* K12) and gram-positive (*S. epidermidis*) by 99.9% [99]. This bifunctional hydrogel holds promise to be applied as a wound dressing and surface coatings for biomedical devices.

1.5 Concluding Remarks

Bacterial surface fouling (biofouling) causes many severe problems in practical applications, such as biomedical field and food industry, which can result in enormous economic losses, increase a large financial burden on the healthcare system, pose serious threats to human health, and even lead to death. Initial bacterial adhesion to surfaces is considered as the first and prerequisite step for biofouling, and protein adsorption on surfaces usually acts as a conditioning film for bacterial adhesion. Therefore, anti-adhesive coatings have been designed as an effective approach to prevent biofouling. Hydrophilic polymers, zwitterionic polymers, superhydrophobic polymers, SLIPS, and protein or glycoprotein-based coatings have exhibited potent anti-adhesive activities. To further improve the anti-biofouling property, bifunctional coatings with anti-adhesive and antibacterial properties have been developed. Anti-adhesive coatings hold great promise to be utilized to prevent biofouling in various fields, whereas they have obvious drawbacks, such as low stability, not suitable for long-term applications, not biodegradable, provoking an immune response, and high cost. Hence, it is still demanding to develop safe, simple, economical, easily available, durable, and effective anti-adhesive coatings in future.

Acknowledgments

This work was supported by grants from Natural Science Foundation of Heilongjiang Province of China (Grant number: LC2018014); Foundation for the Characteristic Discipline of Processing Technology of Plant Foods (Grant numbers: YSTSXK201811 and LTSW201720); Creative Talents Training Program of Heilongjiang Province (Grant number: UNPYSCT-2018105) and Scientific Research Innovation Team Project from Education Department of Heilongjiang Province (Grant number: 135309113) to X. Sun.

References

1. Banerjee, I., Pangule, R.C., Kane, R.S., Antifouling coatings: Recent developments in the design of surfaces that prevent fouling by proteins, bacteria, and marine organisms. *Adv. Mater.*, 23, 690, 2011.
2. Hall-Stoodley, L., Costerton, J.W., Stoodley, P., Bacterial biofilms: From the natural environment to infectious diseases. *Nat. Rev. Microbiol.*, 2, 95, 2004.
3. Song, F., Koo, H., Ren, D., Effects of material properties on bacterial adhesion and biofilm formation. *J. Dent. Res.*, 94, 1027, 2015.
4. Graham, M. and Cady, N., Nano and microscale topographies for the prevention of bacterial surface fouling. *Coatings*, 4, 37, 2014.
5. Lindsay, D. and Von Holy, A., Bacterial biofilms within the clinical setting: What healthcare professionals should know. *J. Hosp. Infect.*, 64, 313, 2006.
6. O'Toole, G., Kaplan, H.B., Kolter, R., Biofilm formation as microbial development. *Annu. Rev. Microbiol.*, 54, 49, 2000.
7. Epstein, A.K., Pokroy, B., Seminara, A., Aizenberg, J., Bacterial biofilm shows persistent resistance to liquid wetting and gas penetration. *P. Natl. A Sci.*, 108, 995, 2011.
8. Helbig, R., Günther, D., Friedrichs, J., Rößler, F., Lasagni, A., Werner, C., The impact of structure dimensions on initial bacterial adhesion. *Biomater. Sci.*, 4, 1074, 2016.
9. Geesey, G.G., Wigglesworth-Cooksey, B., Cooksey, K.E., Influence of calcium and other cations on surface adhesion of bacteria and diatoms: A review. *Biofouling*, 15, 195, 2000.
10. Rana, D. and Matsuura, T., Surface modifications for antifouling membranes. *Chem. Rev.*, 110, 2448, 2010.
11. Hetrick, E.M. and Schoenfisch, M.H., Reducing implant-related infections: Active release strategies. *Chem. Soc. Rev.*, 35, 780, 2006.
12. Pham, V.T.H., Bhadra, C.M., Truong, V.K., Crawford, R.J., Ivanova, E.P., Designing antibacterial surfaces for biomedical implants, in: *Antibacterial surfaces*, p. 89, Springer, Cham, 2015.
13. Beloin, C., Stéphane, R., Ghigo, J.M., Lebeaux, D., Novel approaches to combat bacterial biofilms. *Curr. Opin. Pharmacol.*, 18, 61, 2014.
14. Veerachamy, S., Yarlagadda, T., Manivasagam, G., Yarlagadda, P.K., Bacterial adherence and biofilm formation on medical implants: A review. *P. I. Mech. Eng. H*, 228, 1083, 2014.

15. Darouiche, R.O., Treatment of infections associated with surgical implants. *N. Engl. J. Med.*, 350, 1422, 2004.
16. Brooks, J.D. and Flint, S.H., Biofilms in the food industry: Problems and potential solutions. *Int. J. Food Sci. Tech.*, 43, 2163, 2008.
17. Moreira, J., Fulgêncio, R., Alves, P., Machado, I., Bialuch, I., Melo, L.F., Simões, M., Mergulhão, F.J., Evaluation of SICAN performance for biofouling mitigation in the food industry. *Food Control*, 62, 201, 2016.
18. Newell, D.G., Koopmans, M., Verhoef, L., Duizer, E., Aidara-Kane, A., Sprong, H., Food-borne diseases-the challenges of 20years ago still persist while new ones continue to emerge. *Int. J. Food Microbiol.*, 139, S3, 2010.
19. Srey, S., Jahid, I.K., Ha, S.-D., Biofilm formation in food industries: A food safety concern. *Food Control*, 31, 572, 2013.
20. Endersen, L., O'Mahony, J., Hill, C., Ross, R.P., Coffey, A., Phage therapy in the food industry. *Annu. Rev. Food Sci. T*, 5, 327, 2014.
21. Harding, J.L. and Reynolds, M.M., Combating medical device fouling. *Trends Biotechnol.*, 32, 140, 2014.
22. Lowe, S., O'Brien-Simpson, N.M., Connal, L.A., Antibiofouling polymer interfaces: Poly (ethylene glycol) and other promising candidates. *Polym. Chem-UK*, 6, 198, 2015.
23. Liu, H.W., Chen, C.H., Tsai, C.L., Lin, I.H., Hsiue, G.H., Heterobifunctional poly (ethylene glycol)-tethered bone morphogenetic protein-2-stimulated bone marrow mesenchymal stromal cell differentiation and osteogenesis. *Tissue Eng.*, 13, 1113, 2007.
24. Weimer, B.C., Walsh, M.K., Wang, X., Influence of a poly-ethylene glycol spacer on antigen capture by immobilized antibodies. *J. Biochem. Biophys. Methods*, 45, 211, 2000.
25. Krishnan, S., Weinman, C.J., Ober, C.K., Advances in polymers for anti-biofouling surfaces. *J. Mater. Chem.*, 18, 3405, 2008.
26. Yu, Q., Zhang, Y., Wang, H., Brash, J., Chen, H., Anti-fouling bioactive surfaces. *Acta Biomater.*, 7, 1550, 2011.
27. Zhang, F., Kang, E.T., Neoh, K.G., Huang, W., Modification of gold surface by grafting of poly (ethylene glycol) for reduction in protein adsorption and platelet adhesion. *J. Biomat. Sci.-Polym. E*, 12, 515, 2001.
28. Sharma, S., Popat, K.C., Desai, T.A., Controlling nonspecific protein interactions in silicon biomicrosystems with nanostructured poly (ethylene glycol) films. *Langmuir*, 18, 8728, 2002.
29. Zhang, H. and Chiao, M., Anti-fouling coatings of poly(dimethylsiloxane) devices for biological and biomedical applications. *J. Med. Biol. Eng.*, 35, 143, 2015.
30. Tan, J., McClung, W., Brash, J., Non-fouling biomaterials based on blends of polyethylene oxide copolymers and polyurethane: Simultaneous measurement of platelet adhesion and fibrinogen adsorption from flowing whole blood. *J. Biomat. Sci.-Polym. E*, 24, 497, 2013.

31. Murthy, R., Shell, C.E., Grunlan, M.A., The influence of poly (ethylene oxide) grafting via siloxane tethers on protein adsorption. *Biomaterials*, 30, 2433, 2009.
32. Jeon, S.I. and Andrade, J.D., Protein-surface interactions in the presence of polyethylene oxide: I. Simplified theory. *J. Colloid Interface Sci.*, 142, 149, 1991.
33. Ostuni, E., Chapman, R.G., Holmlin, R.E., Takayama, S., Whitesides, G.M., A survey of structure-property relationships of surfaces that resist the adsorption of protein. *Langmuir*, 17, 5605, 2001.
34. Unsworth, L.D., Sheardown, H., Brash, J.L., Polyethylene oxide surfaces of variable chain density by chemisorption of PEO-thiol on gold: Adsorption of proteins from plasma studied by radiolabelling and immunoblotting. *Biomaterials*, 26, 5927, 2005.
35. Ostuni, E., Chapman, R.G., Liang, M.N., Meluleni, G., Pier, G., Ingber, D.E., Self-assembled monolayers that resist the adsorption of proteins and the adhesion of bacterial and mammalian cells. *Langmuir*, 17, 6336, 2001.
36. Fernandez, I.C.S., Mei, H.C.V.D., Lochhead, M.J., Grainger, D.W., Busscher, H.J., The inhibition of the adhesion of clinically isolated bacterial strains on multi-component cross-linked poly (ethylene glycol)-based polymer coatings. *Biomaterials*, 28, 4105, 2007.
37. Kingshott, P., Wei, J., Bagge-Ravn, D., Gadegaard, N., Gram, L., Covalent attachment of poly (ethylene glycol) to surfaces, critical for reducing bacterial adhesion. *Langmuir*, 19, 6912, 2003.
38. Park, K.D., Kim, Y.S., Han, D.K., Kim, Y.H., Lee, E.H.B., Suh, H., Bacterial adhesion on PEG modified polyurethane surfaces. *Biomaterials*, 19, 851, 1998.
39. Razatos, A., Ong, Y.L., Boulay, F., Elbert, D.L., Hubbell, J.A., Sharma, M.M., Force measurements between bacteria and poly (ethylene glycol)-coated surfaces. *Langmuir*, 16, 9155, 2000.
40. Roosjen, A., Van, D.M.H.C., Busscher, H.J., Norde, W., Microbial adhesion to poly (ethylene oxide) brushes: Influence of polymer chain length and temperature. *Langmuir*, 20, 10949, 2004.
41. Harris, L.G., Tosatti, S., Wieland, M., Textor, M., Richards, R.G., Staphylococcus aureus adhesion to titanium oxide surfaces coated with non-functionalized and peptide-functionalized poly (L-lysine)-grafted-poly (ethylene glycol) copolymers. *Biomaterials*, 25, 4135, 2004.
42. Zeng, G., Ogaki, R., Meyer, R.L., Non-proteinaceous bacterial adhesins challenge the antifouling properties of polymer brush coatings. *Acta Biomater.*, 24, 64, 2015.
43. Duncan, R. and Gaspar, R., Nanomedicine (s) under the microscope. *Mol. Pharm.*, 8, 2101, 2011.
44. Herold, D.A., Keil, K., Bruns, D.E., Oxidation of polyethylene glycols by alcohol dehydrogenase. *Biochem. Pharmacol.*, 38, 73, 1989.

45. Schellekens, H., Hennink, W.E., Brinks, V., The immunogenicity of polyethylene glycol: Facts and fiction. *Pharm. Res.-Dordr.*, 30, 1729, 2013.
46. Xu, L.C. and Siedlecki, C.A., Protein adsorption, platelet adhesion, and bacterial adhesion to polyethylene-glycol-textured polyurethane biomaterial surfaces. *J. Biomed. Mater. Res. B*, 105, 668, 2015.
47. Boulmedais, F., Frisch, B., Etienne, O., Lavalle, P., Picart, C., Ogier, J., Polyelectrolyte multilayer films with pegylated polypeptides as a new type of anti-microbial protection for biomaterials. *Biomaterials*, 25, 2003, 2004.
48. He, T., Jańczewski, D., Guo, S., Man, S.M., Jiang, S., Tan, W.S., Stable pH responsive layer-by-layer assemblies of partially hydrolysed poly (2-ethyl-2-oxazoline) and poly (acrylic acid) for effective prevention of protein, cell and bacteria surface attachment. *Colloids Surf. B*, 161, 269, 2018.
49. Konradi, R., Acikgoz, C., Textor, M., Polyoxazolines for nonfouling surface coatings—A direct comparison to the gold standard PEG. *Macromol. Rapid Commun.*, 33, 1663, 2012.
50. Pasche, S., Paul, S.M.D.P., Vörös, J., Spencer, N.D., Textor, M., Poly (l-lysine)-g raft-poly (ethylene glycol) assembled monolayers on niobium oxide surfaces: A quantitative study of the influence of polymer interfacial architecture on resistance to protein adsorption by ToF-SIMS and *in situ* OWLS. *Langmuir*, 19, 9216, 2003.
51. Pidhatika, B., Moeller, J., Vogel, V., Konradi, R., Nonfouling surface coatings based on poly (2-methyl-2-oxazoline). *CHIMIA*, 62, 264, 2008.
52. Zwaal, R., Comfurius, P., Van Deenen, L., Membrane asymmetry and blood coagulation. *Nature*, 268, 358, 1977.
53. Jiang, S. and Cao, Z., Ultralow-fouling, functionalizable, and hydrolyzable zwitterionic materials and their derivatives for biological applications. *Adv. Mater.*, 22, 920, 2010.
54. Zhang, Y., Liu, Y., Ren, B., Zhang, D., Xie, S., Chang, Y., Zheng, J., Fundamentals and applications of zwitterionic antifouling polymers. *J. Phys. D Appl. Phys.*, 52, 403001, 2019.
55. Ishihara, K., Ziats, N.P., Tierney, B.P., Nakabayashi, N., Anderson, J.M., Protein adsorption from human plasma is reduced on phospholipid polymers. *J. Biomed. Mater. Res.*, 25, 1397, 1991.
56. Yoshimoto, K., Hirase, T., Madsen, J., Armes, S.P., Nagasaki, Y., Non-fouling character of poly [2-(methacryloyloxy) ethyl phosphorylcholine]-modified gold surfaces fabricated by the 'grafting to' method: Comparison of its protein resistance with poly (ethylene glycol)-modified gold surfaces. *Macromol. Rapid Commun.*, 30, 2136, 2010.
57. Yang, W., Xue, H., Li, W., Zhang, J., Jiang, S., Pursuing "zero" protein adsorption of poly (carboxybetaine) from undiluted blood serum and plasma. *Langmuir*, 25, 11911, 2009.
58. Cheng, G., Zhang, Z., Chen, S., Bryers, J.D., Jiang, S., Inhibition of bacterial adhesion and biofilm formation on zwitterionic surfaces. *Biomaterials*, 28, 4192, 2007.

59. Cheng, G., Li, G., Xue, H., Chen, S., Bryers, J.D., Jiang, S., Zwitterionic carboxybetaine polymer surfaces and their resistance to long-term biofilm formation. *Biomaterials*, 30, 5234, 2009.
60. Li, G., Cheng, G., Xue, H., Chen, S., Zhang, F., Jiang, S., Ultra low fouling zwitterionic polymers with a biomimetic adhesive group. *Biomaterials*, 29, 4592, 2008.
61. Hippius, C., Bütün, V., Erel-Goktepe, I., Bacterial anti-adhesive properties of a monolayer of zwitterionic block copolymer micelles. *Mater. Sci. Eng. C*, 41, 354, 2014.
62. Yang, W., Xue, H., Carr, L.R., Wang, J., Jiang, S., Zwitterionic poly (carboxybetaine) hydrogels for glucose biosensors in complex media. *Biosens. Bioelectron.*, 26, 2454, 2011.
63. Mi, L. and Jiang, S., Synchronizing nonfouling and antimicrobial properties in a zwitterionic hydrogel. *Biomaterials*, 33, 8928, 2012.
64. Feng, L., Li, S., Li, Y., Li, H., Zhang, L., Zhai, J., Super-hydrophobic surfaces: From natural to artificial. *Adv. Mater.*, 14, 1857, 2002.
65. Nakajima, A., Fujishima, A., Hashimoto, K., Watanabe, T., Preparation of transparent superhydrophobic boehmite and silica films by sublimation of aluminum acetylacetonate. *Adv. Mater.*, 11, 1365, 1999.
66. Stallard, C.P., Mcdonnell, K.A., Onayemi, O.D., Dowling, D.P., Evaluation of protein adsorption on atmospheric plasma deposited coatings exhibiting superhydrophilic to superhydrophobic properties. *Biointerphases*, 7, 1, 2012.
67. Roach, P., Shirtcliffe, N.J., Farrar, D., Perry, C.C., Quantification of surface-bound proteins by fluorometric assay: Comparison with quartz crystal microbalance and amido black assay. *J. Phys. Chem. B*, 110, 20572, 2006.
68. Koc, Y., De Mello, A.J., Mchale, G., Newton, M.I., Roach, P., Shirtcliffe, N.J., Nano-scale superhydrophobicity: Suppression of protein adsorption and promotion of flow-induced detachment. *Lab Chip*, 8, 582, 2008.
69. Tang, P., Wei, Z., Yan, W., Zhang, B., Hao, W., Lin, C., Effect of superhydrophobic surface of titanium on staphylococcus aureus adhesion. *J. Nanomater.*, 2011, 2, 2011.
70. Privett, B.J., Youn, J., Hong, S.A., Lee, J., Han, J., Shin, J.H., Antibacterial fluorinated silica colloid superhydrophobic surfaces. *Langmuir*, 27, 9597, 2011.
71. Yoon, S.H., Rungraeng, N., Song, W., Jun, S., Superhydrophobic and superhydrophilic nanocomposite coatings for preventing *Escherichia coli* K-12 adhesion on food contact surface. *J. Food Eng.*, 131, 135, 2014.
72. Zhang, X., Wang, L., Levänen, E., Superhydrophobic surfaces for the reduction of bacterial adhesion. *RSC Adv.*, 3, 12003, 2013.
73. Sousa, C., Rodrigues, D., Oliveira, R., Song, W., Mano, J.F., Azeredo, J., Superhydrophobic poly (L-lactic acid) surface as potential bacterial colonization substrate. *AMB Express*, 1, 34, 2011.
74. Yang, W., Zhang, H., Liu, X., Zhou, Z., Slippery liquid-infused substrates: A versatile preparation, unique anti-wetting and drag-reduction effect on water. *J. Mater. Chem. A*, 4, 2524, 2016.

75. Wong, T.S., Kang, S.H., Tang, S.K., Smythe, E.J., Hatton, B.D., Grinthal, A., Bioinspired self-repairing slippery surfaces with pressure-stable omniphobicity. *Nature*, 477, 443, 2011.
76. Xiao, L., Li, J., Mieszkin, S., Fino, A.D., Clare, A.S., Callow, M.E., Slippery liquid-infused porous surfaces showing marine antibiofouling properties. *ACS Appl. Mater. Interfaces*, 5, 10074, 2013.
77. Epstein, A.K., Wong, T.S., Belisle, R.A., Boggs, E.M., Aizenberg, J., Liquid-infused structured surfaces with exceptional anti-biofouling performance. *P. Natl. A Sci.*, 109, 13182, 2012.
78. Rungraeng, N., Yoon, S.H., Li, Y., Jun, S., Development of a self-slippery liquid-infused porous surface (SLIPS) coating using carbon nanotube composite for repelling food debris and microbial biofilms. *T. Asabe*, 58, 861, 2015.
79. Wang, P., Zhang, D., Lu, Z., Slippery liquid-infused porous surface bio-inspired by pitcher plant for marine anti-biofouling application. *Colloids Surf. B*, 136, 240, 2015.
80. Manna, U., Raman, N., Welsh, M.A., Zayas-Gonzalez, Y.M., Blackwell, H.E., Palecek, S.P., Slippery liquid-infused porous surfaces that prevent microbial surface fouling and kill non-adherent pathogens in surrounding media: A controlled release approach. *Adv. Funct. Mater.*, 26, 3599, 2016.
81. Bernbom, N., Jørgensen, R.L., Ng, Y.Y., Meyer, R.L., Kingshott, P., Vejborg, R.M., Bacterial adhesion to stainless steel is reduced by aqueous fish extract coatings. *Biofilms*, 3, 25, 2006.
82. Vejborg, R.M., Bernbom, N., Gram, L., Klemm, P., Anti-adhesive properties of fish tropomyosins. *J. Appl. Microbiol.*, 105, 141, 2008.
83. Cone, R.A., Barrier properties of mucus. *Adv. Drug Deliv. Rev.*, 61, 75, 2009.
84. Bansil, R. and Turner, B.S., Mucin structure, aggregation, physiological functions and biomedical applications. *Curr. Opin. Colloid Interface*, 11, 164, 2006.
85. Crouzier, T., Jang, H., Ahn, J., Stocker, R., Ribbeck, K., Cell patterning with mucin biopolymers. *Biomacromolecules*, 14, 3010, 2013.
86. Bushnak, I.A., Labeed, F.H., Sear, R.P., Keddie, J.L., Adhesion of microorganisms to bovine submaxillary mucin coatings: Effect of coating deposition conditions. *Biofouling*, 26, 387, 2010.
87. Shi, L., Ardehali, R., Caldwell, K.D., Valint, P., Mucin coating on polymeric material surfaces to suppress bacterial adhesion. *Colloids Surf. B*, 17, 229, 2000.
88. Co, J.Y., Crouzier, T., Ribbeck, K., Probing the role of mucin-bound glycans in bacterial repulsion by mucin coatings. *Adv. Mater. Interfaces*, 2, 1500179, 2015.
89. Sun, X.H., Huang, J., Zeng, H.B., Wu, J.P., Protein-resistant property of egg white ovomucin under different pHs and ionic strengths. *J. Agr. Food Chem.*, 66, 11034, 2018.
90. Yu, Q., Wu, Z., Chen, H., Dual-function antibacterial surfaces for biomedical applications. *Acta Biomater.*, 16, 1, 2015.

91. Aumsuwan, N., Danyus, R.C., Heinhorst, S., Urban, M.W., Attachment of ampicillin to expanded poly (tetrafluoroethylene): Surface reactions leading to inhibition of microbial growth. *Biomacromolecules*, 9, 1712, 2008.
92. Aumsuwan, N., Heinhorst, S., Urban, M.W., The effectiveness of antibiotic activity of penicillin attached to expanded poly (tetrafluoroethylene)(ePTFE) surfaces: A quantitative assessment. *Biomacromolecules*, 8, 3525, 2007.
93. Aumsuwan, N., McConnell, M.S., Urban, M.W., Tunable antimicrobial polypropylene surfaces: Simultaneous attachment of penicillin (Gram+) and gentamicin (Gram−). *Biomacromolecules*, 10, 623, 2009.
94. Yuan, S., Wan, D., Liang, B., Pehkonen, S.O., Ting, Y.P., Neoh, K.G., Lysozyme-coupled poly (poly (ethylene glycol) methacrylate)–stainless steel hybrids and their antifouling and antibacterial surfaces. *Langmuir*, 27, 2761, 2011.
95. Gao, G., Lange, D., Hilpert, K., Kindrachuk, J., Zou, Y., Cheng, J.T.J., The biocompatibility and biofilm resistance of implant coatings based on hydrophilic polymer brushes conjugated with antimicrobial peptides. *Biomaterials*, 32, 3899, 2011.
96. Yang, W.J., Cai, T., Neoh, K.G., Kang, E.T., Dickinson, G.H., Teo, L.M., Biomimetic anchors for antifouling and antibacterial polymer brushes on stainless steel. *Langmuir*, 27, 7065, 2011.
97. Follmann, H.D.M., Martins, A.F., Gerola, A.P., Burgo, T.A.L., Nakamura, C.V., Rubira, A.F., Antiadhesive and antibacterial multilayer films via layer-by-layer assembly of TMC/heparin complexes. *Biomacromolecules*, 13, 3711, 2012.
98. Fu, J., Ji, J., Yuan, W., Shen, J., Construction of anti-adhesive and antibacterial multilayer films via layer-by-layer assembly of heparin and chitosan. *Biomaterials*, 26, 6684, 2005.
99. Cheng, G., Xue, H., Li, G., Jiang, S., Integrated antimicrobial and nonfouling hydrogels to inhibit the growth of planktonic bacterial cells and keep the surface clean. *Langmuir*, 26, 10425, 2010.

2

Lignin-Based Adhesives

Xiaolin Luo and Li Shuai*

College of Materials Engineering, Fujian Agriculture and Forestry University, Fuzhou, China

Abstract

The increasing concerns about the depletion of fossil fuels and the environment drive the efforts to search renewable alternative chemicals for sustainable material synthesis. Woody biomass is the most abundant biomass on earth, of which lignin is the second largest component after cellulose. As the most abundant natural phenolic polymer, lignin is a potential substitute of phenol for phenolic resin adhesives synthesis. This chapter focuses on adhesives that were synthesized with renewable lignin. We summarize the sources of industrial lignins and the effects of separation processes on lignin structure and reactivity. The methods for pretreating or modifying technical lignins to improve the reactivity of technical lignins for bio-based adhesives synthesis are discussed and the application performances of adhesives synthesized from modified technical lignins are reviewed. Finally, we present the challenges and prospects of developing adhesives with high lignin substitution ratios and high performances.

Keywords: Lignin, biomass, phenol, formaldehyde, resin, adhesive

2.1 Introduction

Adhesive is a group of polymer-based joint used to bind independently existing substances [1]. It can be simply categorized into fossil- and bio-based adhesives. These adhesives, especially fossil-based adhesives, are mainly used in construction, decoration, papermaking, tape, and stamp industries. The physical properties and chemical composition of adhesives have significant impacts on their processing and application properties [2]. Compared

**Corresponding author*: shuailiuw@gmail.com

Inamuddin, Rajender Boddula, Mohd Imran Ahamed and Abdullah M. Asiri (eds.) Green Adhesives: Preparation, Properties and Applications, (25–56) © 2020 Scrivener Publishing LLC

to bio-based adhesives, fossil-based adhesives have superior properties, such as low processing viscosity, high adhesive strength, good thermal stability, and moisture resistance [3, 4]. As a result, most of the adhesives used in our daily lives are fossil-based resins such as phenol-formaldehyde (PF), urea-formaldehyde (UF), melamine-formaldehyde (MF), and polymeric diphenylmethane diisocyanate (pMDI) resins [5].

Although fossil-based adhesives have good properties, their wide uses have also caused many adverse effects [6]. The depletion of crude oil may finally affect the stable supply of fossil-based chemicals required for synthetic adhesives. Fossil-based adhesives that are hard to be degraded in nature not only pose environmental issues but also significantly affect the recycling of other recycable materials [7], e.g., wood in plywood and cellulosic fibers in paperboard. The formaldehyde, which is widely used in synthesizing adhesives, has been explicitly included in the list of carcinogens. Based on the material safety data sheet (MSDS) of science lab, the acute oral toxicity (LD50) of formaldehyde has been identified as 100 mg/kg (Rat) [3]. Consequently, many governments worldwide have gradually developed restrictive regulations (the California Air Resources Board (CARB) and JAS formaldehyde Rating, etc.) to limit the production and use of fossil-based adhesives [5], especially those containing formaldehyde, while the development of the renewable and environment-friendly adhesives is encouraged [2, 3, 8–12]. It is expected that these challenges could be solved or at least alleviated if bio-based materials can partly or completely replace fossil-based ones to produce industrial adhesives without compromising the processing and application performances.

At present, bio-based materials, such as lignin, starch, vegetable proteins, and oils, have been used by researchers as substitutes for fossil-based chemicals to synthesize bio-based adhesives [2, 3]. Woody biomass is the most abundant biomass on earth, of which lignin is the second largest component after cellulose [13]. In the biosynthesis of woody biomass, lignin itself acts as a glue, which effectively binds cellulose and hemicellulose, increasing the strength of cell wall and the stem of the plant [14, 15]. As the most abundant natural phenolic polymer, lignin is a potential substitute of phenol for phenolic resin adhesives synthesis [16–18]. Although carbohydrates or vegetable oils have their own advantages as raw materials for the production of bio-based adhesives, their uses could harm food safety [11]. Therefore, this chapter mainly focuses on the adhesives that are synthesized from renewable lignin materials. We firstly summarize the sources of industrial lignins and the effects of separation processes on lignin structure and reactivity. The methods for pretreating or modifying technical lignins to improve the reactivity of technical lignins for bio-based adhesives synthesis are further systematically discussed. Moreover, the production methods and application performances of adhesives based on modified technical lignins are also reviewed in detail. Finally, we present the challenges and prospects of developing lignin-based adhesives.

2.2 Native Lignin and Source of Technical Lignin

2.2.1 Native Lignin

About 170 years ago, researchers first isolated lignin from biomass [19, 20]. Subsequently, *p*-coumaryl alcohol, coniferyl alcohol, and sinapyl alcohol were identified as the precursors for lignin biosynthesis (Figure 2.1). These precursors, also called monolignols, are incorporated into lignin polymeric structure as *p*-hydroxyphenyl (H), guaiacyl (G), and syringyl (S) units. Under the catalysis of various enzymes (laccase, peroxidase, and other phenol oxidases) and the combined action of various external factors (light, temperature, humidity, etc.), polymerization of these monolignols occurs via the oxidative radicalization of phenolic hydroxyls followed by the coupling of radicals formed from the delocalization of the phenolic radicals by resonance [21]. Because the phenolic radical can be delocalized through resonance to the positions of 3, 5 (the ortho positions), and/or β positions (Figure 2.1, the structure of native lignin), the coupling of these radicals will form different interunit linkages [22]. A phenolic radical couple with the radical at β or 5 position to forms interunit C-O linkages, i.e., β-O-4 and 4-O-5 linkages, respectively (Figure 2.1). Radicals at the

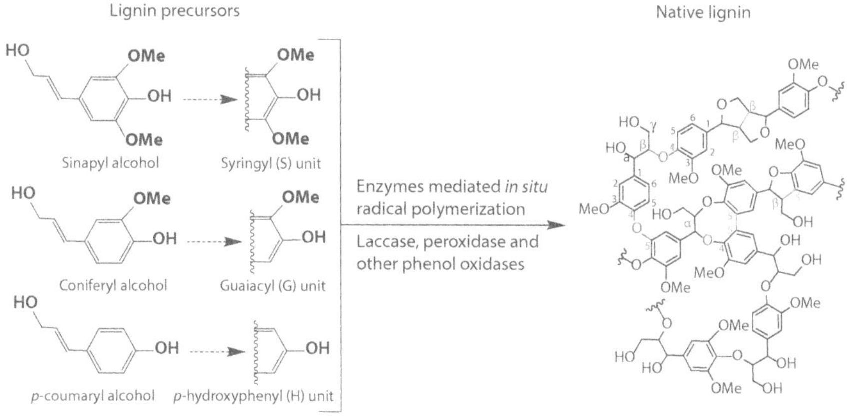

Figure 2.1 Enzymes-mediated *in situ* radical polymerization of monolignols to native lignin.

positions of 5 and β couple with each other form interunit C-C linkages including 5-5, β-5, and β-β linkages (Figure 2.1, the structure of native lignin). In general, softwood lignin mainly consists of G units while hardwood and non-wood lignins are mainly composed of G and S units, and G, S, and H units [23], respectively. The difference of these three units lies in the substitution degree of ortho positions by methoxyls, which is highly associated with the reactivity of lignins as phenol substitutes.

In the process of biological lignification, the interunit bonds formed by the enzyme-mediated radical polymerization/coupling are mainly β-aryl ether bonds [23], which account for 50%–60% of the total linkages in lignin, and the left are mainly C-C linkages [24]. In the process of pulping or biorefinery industries, β-O-4 linkage is typically cleaved to facilitate the removal of lignin with the formation of free phenolic hydroxyls. The amount of free phenolic hydroxyls is another key factor affecting the reactivity of lignin for adhesive synthesis.

2.2.2 Technical Lignins

In the production of pulp or biofuels, the separation of lignin and carbohydrates can be categorized into acid and alkali processes (Figure 2.2).

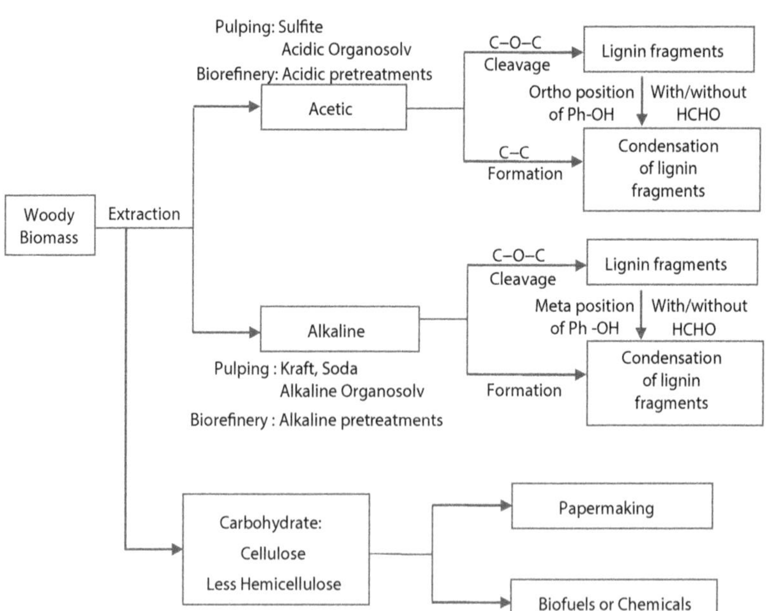

Figure 2.2 Source of technical lignins and effects of extraction processes on its structure.

In the pulping industry, the acidic separation process mainly includes acidic sulfite and organosolv pulping, while the alkali separation process mainly include Kraft, Soda, and alkaline organosolv pulping [25]. The lignins produced by sulfite, Kraft, Soda, and Organosolv pulping are termed as lignosulfonate, Kraft lignin, Soda lignin, and Organosolv lignin, respectively. In the pulping process, lignin is removed and dissolved in the corresponding pulping waste liquor, so-called black liquor or red liquor, upon the cleavage of β-aryl ether bonds. The black liquor or red liquor produced is concentrated by evaporation and burned to recover heat energy, which significantly reduces the value of lignin. Many studies have attempted to separate high-purity lignin from the black liquor and explored its utilization. In the paper and pulping industry, about 50 million tons of lignin is separated as waste every year while only 2% of the lignin waste is further developed as value-added products at present [26]. The successful applications of these lignins include cement additives, dispersants, and surfactants [12].

Other than the above pulping processes, researchers have developed a variety of acidic (such as dilute acid, acidic Organosolv, SPORL, acidic ionic liquids, acidic DES, and acidic hydrotropic pretreatments) and alkaline (such as dilute alkali, alkaline Organosolv, ammonia fiber explosion, and alkaline ionic liquid pretreatments) biorefining methods to separate the three main components (cellulose, hemicellulose, and lignin) in lignocellulosic biomass [27]. In biorefining industry, lignin is mainly recovered from the pretreatment liquor and the solid residue after enzymatic hydrolysis of carbohydrates.

2.3 Limitations of Technical Lignins

2.3.1 Heterogeneity of Technical Lignins

The proportion of the basic structural units of lignin (G, S, and H), and the type and number of linkages vary with biomass types [23]. Even for the same type of biomass, the physicochemical properties of lignin may vary significantly with plantation locations [22]. Therefore, the physicochemical properties (such as molecular weight, thermal stability, and solubility) of the synthesized lignin macromolecules may vary greatly with the source of lignins.

Due to the different lignin depolymerization mechanisms and reaction conditions, the physical properties of technical lignins obtained by typical pulping or biorefining methods are also quite different. For example, the organosolv lignin has a number-average-molecular weight (Mn) of

500–5,000 g/mol, while the molecular weight of lignosulfonates can reach 15,000–50,000 g/mol [28]. The polydispersity (PDI) of molecular weights of different industrial lignins is also different in the corresponding molecular weight range. The organosolv lignin has a PDI of 1.5–2.5, while the PDI of Kraft lignin, lignosulfonate, and Soda lignin is much greater than 2.5 [29]. In addition, other physical properties such as glass transition temperature, solubility in different solvents, and hydrophilicity are also highly different. These variable factors greatly impact lignin quality stability and thereby affect the processability (e.g., viscosity, curing temperature, and time) of lignin-based adhesives.

2.3.2 Reactivity of Technical Lignins

Due to the structural similarities of lignin and phenol, numerous attempts have been made to substitute phenol with lignin for LPF resin synthesis. Unlike the commercialized lignin products (dispersants, surfactants, etc.) that only take advantage of the amphiphilicity of lignin, the production of lignin-based adhesives also requires an understanding of the detailed structure of technical lignins. The incorporation of phenolic hydroxyls into lignin polymer networks leaves a limited number of free phenolic hydroxyls for forming phenolate ions during LPF resin synthesis [30]. Besides, most of the ortho and para positions on the lignin aromatic rings are blocked [31]. Typically, the para position is blocked by the side chain for all units, and some of the ortho positions are blocked by methoxyls as well as 5–5 and β–5 linkages (Figure 2.3a). To initiate the synthesis of PF resins, both free phenolic hydroxyls and free ortho positions are required [28]. In addition, during the removal of lignin, lignin undergoes condensation reactions, which typically involve the consumption of the α-hydroxyls as well as ortho and meta positions (Figure 2.3), causing the further decrease of reactivity.

Technical lignins with a limited number of free phenolic hydroxyls and free ortho positions, therefore, have very low reactivity towards formaldehyde addition and cross-linking [4]. Research work has shown that only an average of 0.3 free ortho positions per lignin unit is available for methylolation or condensation [32], which is only 1/10 of that of a phenol molecule. The ortho and meta positions are further blocked by interunit C-C linkages caused by lignin condensation during biomass pretreatment or pulping under alkaline and acidic conditions (Figure 2.3), respectively. Lignin condensation and lignin depolymerization are parallel reactions [31]. The purpose of the pretreatment process and the pulping process is to partially depolymerize lignin through hydrolyzing β-O-4 linkages and facilitate its

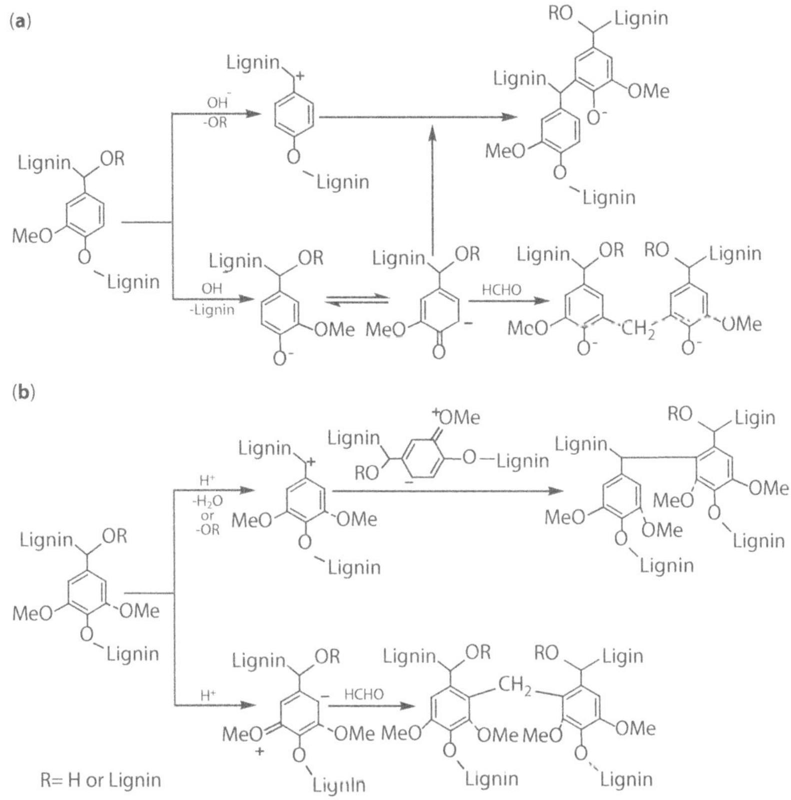

Figure 2.3 Typical condensation reactions of lignin under (a) alkaline and (b) acidic conditions.

removal and dissolution into the pretreatment or cooking solvent [33]. Meanwhile, the α-hydroxyl on the side chain of lignin units, similar to the methylol group of PF resins, can react with the ortho position to form a methylene linkage (Figure 2.3a); in a similar way, formaldehyde generated from the scission of the side chain of lignin during pretreatment or cooking can also lead to the formation of methylene linkages (Figure 2.3b) [31]. These condensation reactions not only lead to repolymerzation (or condensation) of depolymerized fractions into larger molecules but also the consumption of the ortho positions (Figure 2.3b). Besides, condensation may trap some reactive sites in the three-dimensional condensed structure and therefore limit their accessibility to formaldehyde or phenol. A condensed lignin with a high molecular weight and a limited number of reactive sites is not a suitable substitute for phenol or polyols because it may lead to high viscosity, long curing time, and high curing temperature of the resultant lignin-based PF and UF resins as well as weak bonding with the surrounding resins matrix [34–36]. From this perspective, unmodified technical lignin is more of a filler than a substitute for phenols or polyols due to its low reactivity [29]. For example, Jin *et al.* [34] modified PF resin by partially replacing phenol with lignin. The addition of enzymatic hydrolysis lignin by 10 wt% resulted in the highest dry bonding strength of plywood. A few studies [37, 38] could replace phenol with lignin up to 60% without significantly compromising the bonding strength, while some studies [39, 40] rebutted that such high replacement ratios not only deferred the curing process but also harmed other important physical properties (such as water-resistance and morphology) of plywood products.

Overall, lignin has low reactivity towards formaldehyde addition and cross-linking, which could be caused by four reasons. First, during lignin biosynthesis, the phenolic hydroxyl in monolignols is incorporated into lignin to form interunit C-O linkages (etherified phenolic hydroxyls), such as β-O-4 and 4-O-5 linkages, with a limited number of free phenolic hydroxyls available to form phenolate anions which are the activators of FA addition reaction [21]. Second, the ortho positions are essential for FA addition and cross-linking, but most of these positions are blocked by methoxyls [41]. Third, during lignin extraction, lignin condensation will consume some of the ortho positions, further reducing the reactivity of lignin [31]. Fourth, lignin, especially condensed lignin has cross-linked three-dimensional structures, and the accessibility of some of the free ortho positions to phenol or FA could be affected by steric hindrance [24]. Substitution of phenol or polyols with low-reactivity lignin generally leads to inferior mechanical properties, high curing temperatures, and long curing times [2, 10]. Chemical modification of isolated lignins, development of new methods to prevent condensation reaction in the separation process, or a combinational method of the both are proven to be effective ways to improve the utilization value of technical lignins and the quality of lignin-based adhesives.

2.4 Lignin Pre-Treatment/Modification for Adhesive Application

2.4.1 Physical Pre-Treatment

Different botanical sources and separation methods highly affect the physicochemical properties and subsequent processability of technical lignins, which necessitates the development of techniques to obtain the fraction with increased homogeneity, rich functional groups, as well as some other properties. Without chemical input, physical fractionation may be an effective method favored by industry. At present, the methods such as solvent fractionation (organic solvents or anti-solvents), membrane-assisted ultrafiltration, and micro-nano processing have been increasingly used to improve the quality of technical lignins.

1. Solvent fractionation

Depending on the solubility, solvent fractionation can separate the fractions with different physicochemical properties from technical lignins. As early as 1952, Schuerch [42] had investigated the solubility of different lignins (such as Indulin, ethanol organosolv lignin, and native lignins in softwood and hardwood) in different solvents based on Hildebrand solubility parameter and hydrogen bonding theory. The lignin fraction of low molecular weights could easily dissolve in solvents with a wide range of solubility parameters and hydrogen bonding capacity. This pioneering study establishes a solid theoretical foundation for the subsequent solvent fractionation of lignin. Subsequently, researchers attempted to [43] fractionate Alcell® lignin through the extraction of organic solvents of different hydrogen-bonding capacities. Obvious differences between the fractions and the starting lignin were observed in terms of the methoxyl content, fractionation yields, molecular weight distribution, and alkaline nitrobenzene oxidation. The results demonstrated that the fraction of a low molecular weight (average Mw of 714 g/mol) had a lower PDI but a higher content of guaiacyl structures than the fractions of higher molecular weights, favoring the substitution of phenol with lignin for the synthesis of formaldehyde-based resins due to higher content of ortho positions. Also, a technical lignin fraction with high reactive free phenolic hydroxyl groups was also separated through by aqueous acetone evaporation fractionation method [44]. After the lignin cross-linking with polyethylene glycol (PEG) and polymeric methyl-diisocyanate (MDI), the resultant lignin-derived polyurethanes were synthesized from three fractions with distinct molecular weights and chemical functionality [45]. However, the maximum addition of the lignin fraction was limited to only 18 wt%, and beyond that synthesized polyurethanes were too brittle and glassy [45].

To date, different solvent systems have been used to fractionate technical lignins originated from different extraction processes. For example, Sun *et al.* [46] used organic solvents including dichloromethane, methanol, and propanol to fractionate alkaline lignins into four fractions and found that the ratio of non-condensed G to S units in fractions increased considerably with the decrease of their molecular weights. Recently, other organic solvents (i.e., ether, ethyl acetate, acetone, and dioxane/water) were successively developed by this group for overcoming the heterogeneous nature of bamboo organosolv lignin [47]. Similar sequential organic solvent fractionation of technical lignins and/or residual lignins from enzymatic hydrolysis was also explored by other researchers [48–55].

Unlike the sequential solvent extraction, single-step extraction processes for the fractionation of industrial Kraft lignin were developed based on typical (i.e., ethyl acetate, ethanol, methanol, and acetone) [56] and non-typical (i.e., tetrahydrofuran and 2-butanone) [57] organic solvents, resulting in the different Mns and PDIs of the soluble fractions. Anti-solvent-induced precipitation method, e.g., the addition of hexane to acetone-methanol or tetrahydrofuran-methanol co-solvent, was also considered for tuning lignin characteristics [58, 59]. In addition to organic solvents, ionic liquids (ILs) such as 1-buthyl-3-methylimidazolium cation with different anions of chloride ([Bmim]Cl), dimethylphosphate ([Bmim]Me_2PO_4), acetate ([Bmim]OAc), and tosylate ([Bmim]OTs) were also used to fractionate Soda lignin [60], and the purities of the resultant lignin fractions were close to 98%–99%, higher than those resulting from organic solvents. Besides, one study reported that purifying lignin before resin synthesis would be an effective method for improving the thermal stability of lignin-based PF resin adhesive [39].

2. Ultrafiltration

Currently, membrane-assisted ultrafiltration was gradually involved in the process of technical lignin fractionation. Membrane-assisted ultrafiltration was effective to selectively control the molar mass distribution of the isolated fractions and other corresponding properties depending on the cuts of membranes [61]. With three membrane cut-offs of 5, 15, and 50 kDa, Costa *et al.* [62] reported that the fraction passed through 5 kDa membrane possessed the highest content of uncondensed structures according to nitrobenzene oxidation assays and ^{13}C and ^{31}P NMR characterizations. The combination of organic solvent extraction and membrane ultrafiltration was further verified as an effective method to selectively control the molecular mass distribution and properties of lignin fractions [63]. A comparative study using solvent fractionation, successive precipitation, and ultrafiltration demonstrated that the key properties (such as molecular weight, PDI, and functional groups) of isolated lignin fractions could be tailored for different end uses [64]. After the fractionation of ammonium-based sulfite spent liquor by membrane ultrafiltration, a high-quality wafer board could be produced with phenolic resin adhesive synthesized from low-molecular-weight lignosulfonate, while the use of high-molecular-weight lignosulfonate resulted in poor mechanical properties [65].

3. Nano-processing

Processing lignin into nanoparticles can also improve its processability and application properties. Nano-lignin particles with controllable sizes were simply produced by diluting the hydrotropic pretreatment (*p*-toluenesulfonic acid and water) liquor with water followed by centrifugation with different centrifugal forces [66]. Lignin nanoparticles can significantly add value to the applications of lignin from pulping or biorefinery industry [67]. The adhesiveness and toughness of lignin-based hydrogel synthesized from lignin nanoparticles was highly improved [68]. The nano-treatment exposed more active functional groups to the surface of lignin and thereby reduced the steric hindrance of technical lignin to small molecular compounds (i.e., formaldehyde and phenol). Screw extrusion was found to change the microstructure of sorghum lignin (SL) [69]. After the extrusion, the intensity of FT-IR absorption peak for free hydroxyls was increased as compared to that without pre-treatment, and the shear strength and water resistance of wood veneer joints bonded by modified lignin-soy protein adhesives were thus obviously improved [69].

2.4.2 Chemical Modification

In addition to the physical pretreatments, chemical modification methods are also important for overcoming the structure-mediated low reactivity of technical lignins caused by blocked ortho and para reactive sites on the phenyl group [31] and the shortage of aliphatic and phenolic hydroxyl groups [70]. Without modification, lignin is only added as a filler to adhesives and other composite materials, which not only affects subsequent processing efficiencies (such as curing temperature and time), but may also sacrifice product performances [5]. Therefore, various chemical modification methods (such as demethylation, phenolation, hydroxymethylation, amination, nitration, etherification, and esterification) have been developed to introduce new reaction sites to lignin macromolecules to improve their reactivity towards the synthesis of bio-based adhesives [28, 71].

(1) Demethylation

In the process of biosynthesis, the ortho positions in the lignin G or S unit is usually substituted by one or two methoxy groups (Figure 2.1). In addition, the phenolic hydroxyl groups of the lignin units (G, S, or H) are etherified (Figure 2.1). As a result, the reactivity of lignin is greatly suppressed due to the absence of free phenolic hydroxyls [32]. Therefore, the removal of the methyl group from lignin will facilitate the release of free phenolic hydroxyl groups for synthesizing lignin-based bio-adhesives (such as UF and MF) [10].

Under alkaline conditions, molten sulfur-mediated lignin demethylation (Figure 2.4a) has been industrialized by Gaylord Chemical Company (US) [28, 71]. During this process, this method could be used to remove methyls from Kraft and Soda lignins [72, 73], leading to a double or more increase of the phenolic hydroxyl group. When 60% of the phenol was replaced by the demethylated lignin, the performance of plywood made with the synthesized lignin-phenol-formaldehyde (LPF) adhesive reached Chinese national standard for first-class plywood. With the same demethylation method, Liu and Li [74] developed another wood adhesive based on the cross-linking between demethylated Kraft lignin (DKL) and polyethylenimine (PEI). Lap-shear specimens using DKL-PEI adhesive exhibited high shear strength and water resistance. With increased phenolic hydroxyl groups, the curing mechanism of such new adhesive was similar to that of quinone-tanning process in nature. Recent studies reported that Na_2SO_3 could release more phenolic hydroxyl groups from agricultural crop lignin than other three reagents (S, NaSH, and *n*-dodecyl mercaptan), resulting in improved bond strength (0.92 to 1.07 MPa) and lower

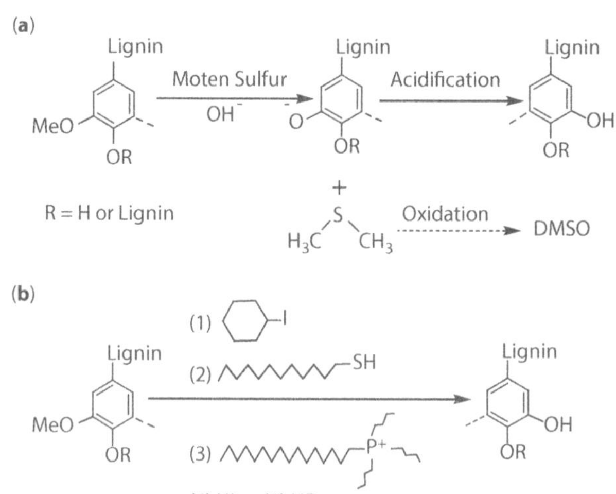

Figure 2.4 Demethylation of lignin fragments by different reactants under (a) basic and (b) acidic conditions.

formaldehyde emissions (0.22 to 0.58 mg/L) of plywoods bonded with the modified LPF [75, 76].

Under acidic conditions, demethylation of lignin can be effectively conducted with other reagents such as 1-dodecanethiol, iodocyclohexane, hexadecyltributylphosphonium bromide, hydroiodic, and hydrobromic acid (Figure 2.4b). For example, softwood Kraft lignin (SKL) was demethylated by iodocyclohexane with dimethyl formamide (DMF) as a solvent in a reflux reactor at 145°C for 3 h [77]. The total phenolic hydroxyl group increased from 1.09 mmol/g for non-modification sample to 1.75 mmol/g for modified SKL, while the molecular weight of this lignin decreased after demethylation. Similarly, when iodocyclohexane was used as the demethylation reagent, the PDI of the lignin molecular mass was significantly reduced and the phenolic hydroxyl content was increased by three times [78]. The wet and dry bonding strengths of resultant Lap-shear wood samples were also highly stronger than that of the sample using unmodified LPF. Lignin-based epoxy resin was synthesized using demethylated enzymatic hydrolysis lignin (DML) that used 1-dodecanethiol as a nucleophilic reagent and sodium methoxide as a catalyst [79]. The modified lignin-based epoxy resin possessed better elasticity than that of unmodified lignin-based product upon curing. Sawamura *et al.* [80] confirmed that iodocyclohexane or hydroiodic acid was more effective than 1-dodecanethiol for removing methyl from technical lignins. These results indicate that demethylation could be a good method to enhance the reactivity of lignin towards bio-adhesives synthesis. Despite the remarkable improvements observed in the mechanical performances of synthesized resins, the curing temperature and/or time of demethylated lignin-based adhesives for engineered wood panels application is still inferior to conventional phenolic adhesives due to the inherent disadvantages such as the low ortho position availability and steric hindrance. Therefore, it is highly desirable to develop more effective chemical modifications or other methods to enhance the reactivity of lignin.

(2) Phenolation

The pre-reaction of lignin with phenol or its derivatives is commonly referred as phenolation or phenololysis, which is one of the most widely chemical modification methods for increasing the active reaction sites in lignin, particularly phenyl ortho and/or para positions [81].

In acidic media, the electron of aromatic ring is dictated by phenolic hydroxyl via resonance and/or inductive effects, resulting in higher electron density at the ortho and para positions of phenolic hydroxyl (Figure 2.5a) [31]. In addition, the benzylic cation can form in the lignin

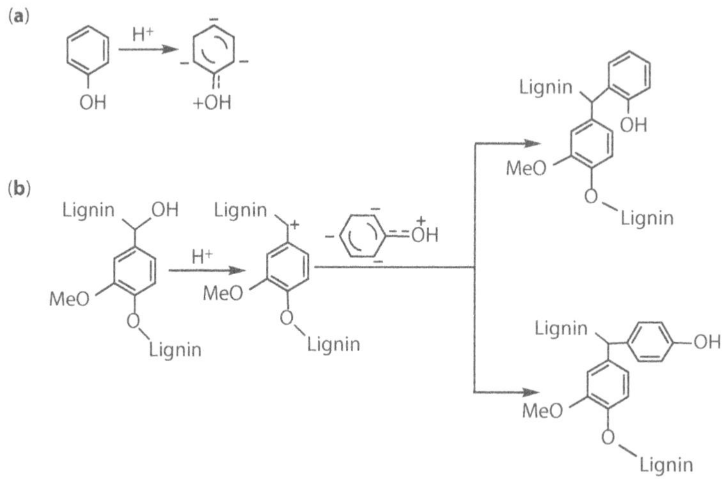

Figure 2.5 (a) Resonance structures of phenol and (b) phenolation at the side chain (Cα position) of lignin with phenol through a nucleophilic reaction under acidic condition.

macromolecular structure through an acid-catalyzed dehydration reaction (Figure 2.5b) [82]. Therefore, Funaoka *et al.* proposed that the ortho and para positions of phenol could preferably react with lignin at Cα-positions of side chains through a nucleophilic reaction (Figure 2.5b) [83]. This reaction pathway was further confirmed with a lignin model compound (guaiacylglycerol-β-guaiacyl ether, GG) [84] and real lignin [85]. Besides, some researchers also speculated that a phenolation reaction can also occur on the aromatic nuclei of lignin other than the side chains [86, 87].

Subsequently, studies on the modification of lignin via phenolation in acidic media were reported successively. By using oxalic acid as a catalyst, a softwood lignosulfonate was successfully phenolated at 120°C for 160 min [84]. ^{1}H NMR analysis indicated that the phenolation reaction preferred to occur at the ortho position of phenol over its para position. Subsequently, a PF resin was synthesized though partially replacing phenol with phenolated rice hull lignin followed by a three-step polymerization method and achieved better water and ultraviolet resistance as compared to unmodified lignin [88]. For a technical alkali lignin, the number of phenolic hydroxyls increased by nearly three times after phenolation [89]. With 10% substitution of phenol with lignin, Lee *et al.* [90] found that phenolated organosolv oak wood lignin-based PF (POLPF) resin had comparable bond strength to conventional PF adhesives. However, they also pointed out that the severe separation conditions of lignin were not conducive to improving the performance of lignin-based adhesives through phenolation modification because of the adverse effect of condensation reactions on the structure of lignin itself. For an enzymatic hydrolysis lignin (EHL), the condensation of lignin during extraction process was not favorable to the synthesis of bio-adhesives used for plywoods [85]. Podschun *et al.* [91] reported that the free ortho and para positions (9.0 mmol/g) of phenolated beech organosolv lignin was much higher than that of unmodified sample (0.7 mmol/g) according to NMR analysis. They contended that organosolv lignin with low MWs and PDIs benefited the properties of lignin-based thermosetting resins. They further reported that hardwood organosolv and softwood sulfite lignins were more reactive towards phenolation-induced activation than those of grass Soda and softwood Kraft lignins due to the presence of more aliphatic hydroxyl groups [92]. With a phenol substitution level of 20% or 40%, organosolv and sulfite lignins were efficiently incorporated into PF resins [93]. The wet and dry internal bond strengths of the resulting particleboards met the European standards. Similar results reported by other groups further verified these findings [86, 94].

Without fully understanding the reaction mechanism, Sun's group found that the phenolation of lignin could also be effectively processed under alkaline conditions. The active reaction sites of four biorefinery lignins from different feedstocks were effectively increased by phenolation modification and could replace 50% of phenol for synthesizing LPF resin adhesives [95]. The LPF-derived bonding strength and the emissions of formaldehyde of three-layer plywood fully met Chinese National Standards of the first grade plywood. However, the conditions required for the curing process, especially the curing temperature, were harsher than those for the traditional PF resin adhesives [35].

Besides, lignin-rich biorefinery residues had also been considered as substitutes of phenol for adhesives preparation in acidic or alkaline media. Lignocellulosic butanol residue (BR)-based PF resin was synthesized and showed acceptable mechanical and thermal properties when the maximal phenol substitution was 30% [96]. Another LPF resin adhesive was also prepared based on phenolated steam explosion lignin (SEL) [97]. With high phenol substitution, the resultant properties of plywoods reached the standard of first class plywood in China. Recently, bio-crude obtained from the liquefaction of Fir sawdust with phenol as a solvent under an acidic condition at 135°C–165°C was modified by urea at high temperature (>130°C). A new thermosetting phenolic resin was synthesized with phenol-urea-modified bio-crude and paraformaldehyde. The mechanical performances and thermal stability of the resultant material were close or even superior to those of conventional phenolic laminate but curing temperatures as high as 185°C–220°C were needed. Wheat straw lignin residue after an autohydrolysis refining followed by enzymatic hydrolysis process was phenolated to improve its reactivity and enabled a higher phenol substitution in adhesives application [87].

In addition to LPF resin, phenolated lignin or its derivatives can also be used to prepare other adhesives under acidic or alkaline conditions. Phenolated Kraft lignins (PKL) were used to synthesize UF resin for engineered wood panels production [98]. When the substitution of urea with PKL was lower than 20%, the internal bond strength of the panels bonded with the PKL-based UF resin was comparable to that of the unmodified UF resin while less formaldehyde emission was detected [98]. A bio-based novolac resin cross-linked by hexamethylenetetramine (HMTA) was prepared using phenolated enzymatic hydrolysis lignin and hydroxymethylfurfural (HMF) [99]. Although the thermal stability was improved, the cured tensile strength (89 MPa) of this lignin-based novolac resin was still lower than that of the fossil-based phenol-HMF resin (110 Mpa) by about 10% [99]. Zhao *et al.* [100–101] recently prepared two renewable epoxy thermosets through co-oligomerizing phenolated lignin and lignin-derived monomer (dihydroeugenol or salicyl alcohol) and resulted in thermomechanical properties comparable to those of the conventional bisphenol A-based counterpart [101]. Another lignin-based epoxy resin was also synthesized by reacting phenolated lignin with epichlorohydrin. With 16% content of phenolated lignin in petroleum-based epoxy precursors, phenolated lignin-based epoxy resin showed better tensile and heat-resistance properties as compared to unmodified lignin, but unfortunately, the curing performance was not effectively improved [102].

Based on the modification of technical or biorefinery lignins via a phenolation step, higher phenol substitution and better application properties of modified lignin-based adhesives (such as PF, UF, and epoxy resins) were achieved by different research groups. However, without compromising the physicochemical properties of the resultant resins-based adhesives, there are still two challenges that need to be further studied. On the one hand, fossil-based phenol cannot be fully replaced, and the highest substitution ratio reported so far is still less than 50%; on the other hand, because the chemical modification cannot completely change the complex and heterogeneous structural characteristics of lignin, more severe curing conditions are needed for plywood production if excessive lignin is incorporated into the matrix of traditional thermoplastic adhesives.

(3) Methylolation

PF resins are synthesized through the reaction of formaldehyde with phenol. Depending on the molar ratio of formaldehyde-to-phenol (F/P), two types of resins called Resole (Figure 2.6a, F/P ratio > 1, alkaline media) and novolac (Figure 2.6b, F/P ratio < 1, acidic media) will form [4]. For these two PF resins, formed methylols will directly participate in the curing process to form methylene linkages. This structure is equally important if lignin is used in place of phenol to synthesize LPF. Although lignin can react with formaldehyde under alkaline or acidic conditions, formaldehyde is more prone to react first with phenol if lignin is directly mixed in a lignin-phenol-formaldehyde mixture due to the predominantly free ortho and para positions of phenol. Therefore, methylolation modification is often adopted to pre-introduce methylol groups into technical lignin prior to the synthesis of LPF resins.

Under alkali conditions, deprotonation of a lignin molecule forms a phenolate anion which is delocalized to the ortho positions by resonance, and these activated ortho positions can be attacked by FA molecules to form methylol (or hydroxymethyl) groups (Figure 2.7b). Under

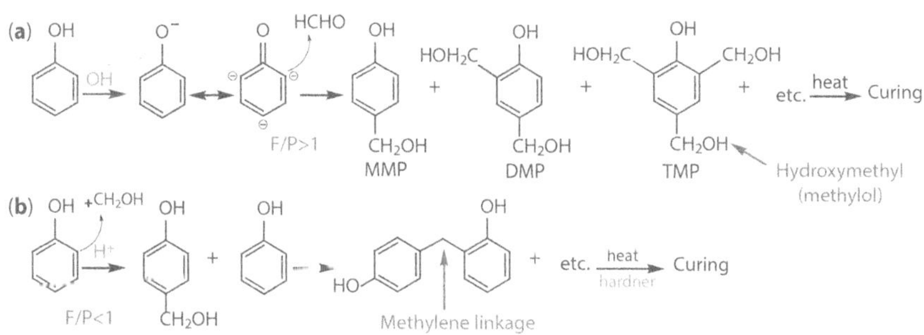

Figure 2.6 The synthetic mechanisms of (a) resole resin and (b) novola resin.

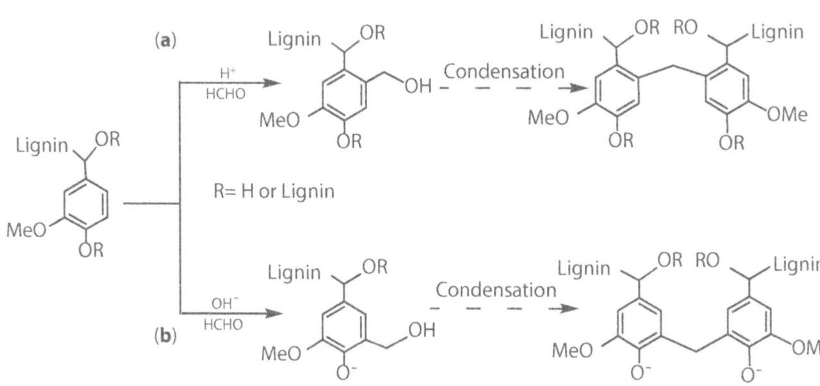

Figure 2.7 Methylolation modification of lignin under (a) acidic and (b) alkaline conditions.

acidic conditions, methylol groups form at the meta positions of phenolic hydroxyl group (Figure 2.7a). The amount of ortho and meta positions of real lignin is a key factor affecting the efficiency of lignin methylolation modification. Moreover, two side reactions are often accompanied: (1) methylol groups may also cause the condensation of lignin itself [31, 82]; (2) formaldehyde may also undergo self-condensation via well-known Canizzaro reaction [28, 71]. Thus, the process conditions (F/P ratio, solvent, the amount of catalysts (i.e., acid or base), temperature, time, etc.) of methylolation modification as well as lignin structural characteristics should be highly optimized.

By varying modification process conditions, different methylolated lignin-based adhesives were successively produced. The ratio of formaldehyde to methylolated Formacell eucalyptus lignin was important for resols synthesis [103]. This group subsequently found that the molecular weight of hydroxymethylated Organosolv lignins highly affected the apparent activation energies of LPF during curing process [104]. For a methylolated softwood ammonium lignosulfonate, the optimal ratios of formaldehyde to modified lignin (ML) and the substitution of phenol with ML were 2.5% and 30%, respectively; the properties of plywood bonded with the synthesized LPF resin could meet the requirements [105]. From the perspective of the reaction mechanism, high syringyl-to-guaiacyl unit ratio and total methoxy group content could be the most unfavorable factor leading to the poor reactivity of bamboo alkaline lignin towards methylolation rather than the molecular weight and PDI [106].

Moreover, based on methylolation modification, about 0.36 mole of the $-CH_2OH/C_9$ unit was introduced into a pine Kraft lignin [107]. Under the same curing condition (210°C, 8 min, resin 5 wt%), a bonding strength of about 65 psi was obtained by mixing hydroxymethylated Kraft lignin (HMKL) and PF resin with a ratio of 1:1. However, the researchers further mentioned that decreasing the addition of HMKL favored the bonding strength. When the substitution of phenol with methylolated acetic acid eucalyptus lignin was less than 40%, the quality of pine and eucalyptus plywood boards bonded by modified lignin-based PF resin adhesive fulfilled the European Standard EN 314-1:1993 for wood-based panel (WBP) production [108]. Recently, liquefied bio-oils from pine sawdust was also methylolated with formaldehyde under alkaline conditions for the production of resol resins [109]. With a phenol substitution up to 60%, the mechanical strength of plywood specimen was comparable to that using the conventional PF resol as an adhesive [110]. The advantages of high homogeneity, low molecular weight, and more reactive sites of such bio-oils may contribute to this high phenol substitution as compared to technical lignins.

(4) Other modification methods

In addition to above widely used modification methods, etherification, pyrolysis, and hydrolysis had also been gradually developed to activate lignin for synthesizing adhesives. The effects of these modification methods on the properties of corresponding lignin-based adhesives were thereby discussed as follows.

It is well known that polyurethane-based adhesives (PUAs) can be effectively synthesized through the reaction between diisocyanates (Figure 2.8a) and polyols [10]. Due to the presence of various hydroxyl groups, lignin can react with diisocyanate to form PU-based adhesive (Figure 2.8b). In the PU synthesis process, the researchers found that the reaction rate of aliphatic alcohols (including primary and secondary alcohols) with diisocyanates was much higher than that of phenolic hydroxyl groups [111]. However, in the biosynthesis process, the lignin side chain hydroxyl groups are usually linked by other functional groups in the form of aryl ether bonds (Figure 2.1) [14, 15, 23]; in addition, the residual side chain hydroxyl groups are partially removed or trapped in condensed structure during the lignin separation process [13, 27, 31, 33]. Etherification would be a good method to create new aliphatic hydroxyl and thereby improve the reactivity of lignin for PU-based adhesives synthesis.

Propylene oxide- or propylene carbonate-based etherification has been developed for transferring lignin phenolic hydroxyl group to a more reactive aliphatic hydroxyl group (Figure 2.9) [10], which facilitates the cross-link between diisocyanates and lignin as a polyol. Compared to the

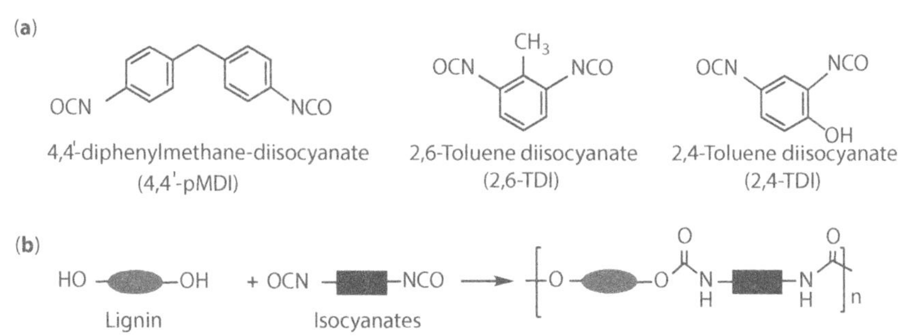

Figure 2.8 (a) Common molecular structure of diisocyanates. (b) The cross-linking reaction between lignin and diisocyanates.

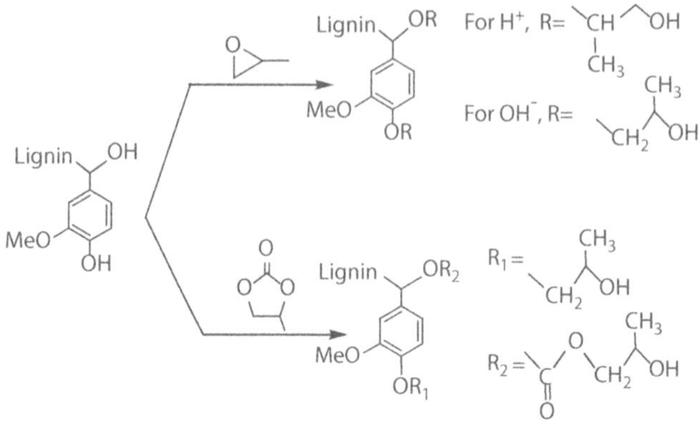

Figure 2.9 Etherification of lignin phenolic hydroxyl group with propylene oxide and propylene carbonate.

unmodified lignin, the reactivity of etherified steam-explosion straw lignin towards diisocyanate was improved [112]. With the etherified lignin, 4,4'-pMDI as polyol and diisocyanate, the resultant PU-based adhesive demonstrated higher shear strength than that using the unmodified lignin [113]. When 30% of fossil-based polyols was replaced by hydroxypropylated lignin, the bonding strength of Southern yellow pine particle board was higher than that of the conventional PU-based adhesive. However, another study claimed that the incorporation of more lignin into PU-based adhesives would cause brittleness and phase separation phenomenon [114].

Pyrolysis has been widely investigated to convert biomass to bio-oils which can be further fractionated and upgraded into fuels and chemicals [115]. Pyrolysis of lignin generates a decent yield of lignin oils composed of small lignin fragments, but the formation of unknown compounds and coke is the major issue of this method. A typical weight distribution of lignin pyrolysis products at 450°C–550°C is: 55% char, 35% liquids, and 12% gases [116]. Further extraction and purification of useful phenolic compounds from bio-oil for LPF resin synthesis are required [117].

Another way to enhance lignin reactivity is to hydrolyze lignin into small fractions to generate more phenolic hydroxyl groups so that existing free ortho positions can become reactive towards formaldehyde addition. Similar to the hydrolysis of β-O-4 linkages occurring during biomass pretreatment or pulping cooking, the residual β-O-4 linkages in technical lignin (referring to lignin separated from biomass with chemical treatments) can be further disrupted to free additional phenolic hydroxyls under acidic or alkali conditions. The number of phenolic hydroxyls increased after lignin hydrolysis [118]; however, lignin will further condense to form more C-C linkages under acidic or alkali conditions, thereby consuming the ortho positions and α-hydroxyls and reversely reducing the reactivity of lignin [119]. Therefore, the efforts to increase the number of free phenolic hydroxyls at the sacrifice of the free ortho positions will not increase the reactivity of lignin because both the free phenolic hydroxyl and the free ortho position are essential for formaldehyde addition and cross-linking.

2.5 Challenges and Prospects

Because of the increasing concerns about the depletion of fossil fuels and the environment, searching for renewable alternative chemicals for sustainable material synthesis has gained a lot of attention these years [120–123]. Phenol and/or polyol as an important commodity chemicals are also the targeted chemicals. In order to improve phenol or polyols substitution without compromising the properties of resultant resin-based adhesives, various chemical modifications mainly including demethylation (Figure 2.4), phenolation (Figure 2.5), methylolation (Figure 2.6), and etherification (Figure 2.7) have been conducted to increase the reactivity of lignin. However, the substitution ratio of fossil-derived platform chemicals such as phenol with ML is still limited to a bottleneck value of 50% [39, 71]. Over 50% substitution of phenol or polyols compromises the properties of resultant resins [34, 36]. This is due to the limited reactive sites and steric hindrance of lignin macromolecules.

Studies have shown that for every lignin unit of Kraft lignin, only 0.3 free ortho positions are available for FA addition and cross-linking [32]; additionally, most of these positions are not activated due to the lack of phenolic hydroxyls [124]. When lignin is consider as a filler which is added to PF resin for reinforcement (Figure 2.4), methylolation and phenolation modification should not be considered effective to increase the number of reactive sites on the surface of lignin because their binding points with the PF matrix have already been defined by its own structure, as depicted in Figure 2.10a. From the structure point of view, there is no difference between the resins synthesized with untreated and methylolation- and phenolation-modified lignin. Specifically, phenolation increases the reactive sites of lignin through grafting three-reactive-site phenol onto lignin, but the number of grafted phenol molecules is limited by the number of α-hydroxyls (or methylol-like moieties) available in a lignin molecule [91]; similarly, methylolation degree depends on the availability the free ortho

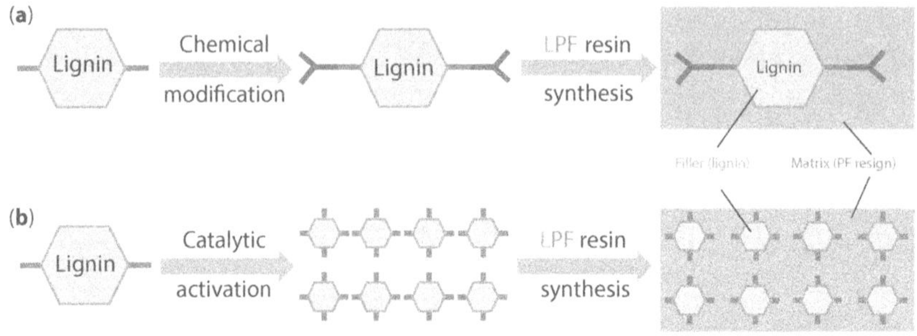

Figure 2.10 Schematic representations of (a) chemical modification and (b) catalytic activation of lignin for lignin-based PF (LPF) resin synthesis.

positions in lignin. The major advantage of methylolation and phenolation modification is that lignin is pretreated with FA or phenol in advance in a low-viscosity environment to ensure that all available reactive sites are fully utilized. Without these pretreatment modifications, high-reactivity phenol will compete with low-reactivity lignin towards FA addition and condensation, which will harm the incorporation of lignin into the network of resins. Pyrolysis of lignin generates unknown compounds and coke, which decrease carbon efficiency and require purification steps to extract useful lignin fragments [117]. Hydrolysis generates free phenolic hydroxyls through partially breaking β-O-4 linkages, but the parallel condensation reaction leads to the increase of lignin molecular weight and further consumption of the ortho positions [119].

In summary, all of the present modification methods cannot create new active sites at the ortho position. A potential strategy could be a combinational method of depolymerizaiton and demethoxylation. In recent years, hydrogenolytic depolymerization of lignin into monomers has been considered as a promising pathway to valorize lignin [125, 126]. Hydrogenolysis will also remove functional groups, such as C = O and C = C bonds, which cause the dark color of lignin, and generate light-colored lignin products for synthesizing high-quality resins. This is a unique feature of hydrogenolysis because existing modification methods cannot change the dark color of lignin. Other than cleaving all β-O-4 linkages to generate free phenolic hydroxyls, hydrogenolysis can simultaneously remove methoxyls from the lignin aromatic ring [127]. Depolymerization and demethoxylation will increase the number of free ortho positions per lignin units, therefore increasing the binding of activated lignin products with PF matrix (Figure 2.10b). When lignin is further depolymerized to small molecules similar to phenol (this can be considered as the extreme case of Figure 2.10b, in which lignin and PF matrix become miscible and a homogenous phase), the molecules are expected to be directly used to synthesize resins without the addition of phenol.

2.6 Conclusions

From the perspective of circular economy, high value-added uses of lignin can create profitable economy for the pulping and biorefinery industries. Because of its aromatic and polyhydroxyl structure, lignin has the great potential as a substitute of fossil-based phenol or polyol for the synthesis of adhesives, which will reduce excessive dependence on fossil resources. However, the heterogeneity of lignin itself and the adverse effects of the separation process on its important structural functional groups limit the application properties of the lignin-based adhesives. In the process of adhesive synthesis, although researchers have developed many physical pre-treatments (such as solvent fractionation and ultrafiltration) and chemical modification methods (such as demethylation, phenolation, and methylolation) to improve the reactivity of lignin, the application of modified lignin-based adhesives is still not comparable to conventional fossil-based adhesives (such as PF and PU) due to the limited reactive sites of lignin itself. In particular, the remaining defects of modified lignin-based adhesives are mainly reflected in the following two aspects: (1) the substitution ratio of fossil-based phenol or polyol with lignin is still very limited; (2) high substitution ratios cause more severe curing conditions and/or worsening properties of plywood. A potential strategy of solving this issue is the depolymerization of lignin followed by demethoxylation to increase the amount of both free phenolic hydroxyls and ortho positions for lignin-phenol-formaldehyde resin synthesis.

References

1. Pizzi, A., *Advanced wood adhesives technology*, CRC Press, Boca Raton, 1994.
2. Heinrich, L.A., Future opportunities for bio-based adhesives – advantages beyond renewability. *Green Chem.*, 21, 1866–1888, 2019.
3. Ferdosian, F., Pan, Z., Gao, G., Zhao, B., Bio-Based Adhesives and evaluation for wood composites application. *Polymers (Basel)*, 9, 70, 2017.
4. Pilato, L., Phenolic resins: 100Years and still going strong. *React. Funct. Polym.*, 73, 270–277, 2013.

5. Ang, A.F., Ashaari, Z., Lee, S.H., Md Tahir, P., Halis, R., Lignin-based copolymer adhesives for composite wood panels – A review. *Int. J. Adhes. Adhes.*, 95, 102408, 2019.
6. Pizzi, A., Recent developments in eco-efficient bio-based adhesives for wood bonding: Opportunities and issues. *J. Adhes. Sci. Technol.*, 20, 829–846, 2006.
7. Imam, S.H., Bilbao-Sainz, C., Chiou, B.-S., Glenn, G.M., Orts, W.J., Biobased adhesives, gums, emulsions, and binders: Current trends and future prospects. *J. Adhes. Sci. Technol.*, 27, 1972–1997, 2013.
8. Hemmilä, V., Adamopoulos, S., Karlsson, O., Kumar, A., Development of sustainable bio-adhesives for engineered wood panels – A Review. *RSC Adv.*, 7, 38604–38630, 2017.
9. Kumar, S., Samal, S.K., Mohanty, S., Nayak, S.K., Recent development of biobased epoxy resins: A review. *Polym.-Plast. Technol. Eng.*, 57, 133–155, 2018.
10. Alinejad, M., Henry, C., Nikafshar, S., Gondaliya, A., Bagheri, S., Chen, N., Singh, S.K., Hodge, D.B., Nejad, M., Lignin-Based Polyurethanes: Opportunities for Bio-Based Foams, Elastomers, Coatings and Adhesives. *Polymers (Basel)*, 11, 1202, 2019.
11. Tribot, A., Amer, G., Abdou Alio, M., de Baynast, H., Delattre, C., Pons, A., Mathias, J.-D., Callois, J.-M., Vial, C., Michaud, P., Dussap, C.-G., Wood-lignin: Supply, extraction processes and use as bio-based material. *Eur. Polym. J.*, 112, 228–240, 2019.
12. Wang, H., Pu, Y., Ragauskas, A., Yang, B., From lignin to valuable products-strategies, challenges, and prospects. *Bioresour. Technol.*, 271, 449–461, 2019.
13. Karimi, K. and Taherzadeh, M.J., A critical review of analytical methods in pretreatment of lignocelluloses: Composition, imaging, and crystallinity. *Bioresour. Technol.*, 200, 1008–1018, 2016.
14. Ralph, J., Brunow, G., Harris, P.J., Dixon, R.A., Schatz, P.F., Boerjan, W., Lignification: Are lignins biosynthesized via simple combinatorial chemistry or via proteinaceous control and template replication?, in: *Recent advances in polyphenol research*, vol. 1, F. Daayf and V. Lattanzio, (Eds.), pp. 36–58, Wiley-Blackwell, Oxford, 2008.
15. Reale, S., Attanasio, F., Spreti, N., De Angelis, F., Lignin chemistry: Biosynthetic study and structural characterisation of coniferyl alcohol oligomers formed *in vitro* in a micellar environment. *Chem.-Eur. J.*, 16, 6077–6087, 2010.
16. Solt, P., Rossiger, B., Konnerth, J., van Herwijnen, H.W.G., Lignin phenol formaldehyde resoles using base-catalysed depolymerized kraft lignin. *Polymers*, 10, 1162, 2018.
17. Solt, P., van Herwijnen, H.W.G., Konnerth, J., Thermoplastic and moisture-dependent behavior of lignin phenol formaldehyde resins. *J. Appl. Polym. Sci.*, 136, 48011, 2019.

18. Lettner, M., Solt, P., Rossiger, B., Pufky-Heinrich, D., Jaaskelainen, A.S., Schwarzbauer, P., Hesser, F., From wood to resin-identifying sustainability levers through hotspotting lignin valorisation pathways. *Sustainability*, 10, 2745, 2018.
19. Phillips, M., The chemistry of lignin. *Chem. Rev.*, 14, 103–170, 1934.
20. Adler, E., Lignin chemistry—Past, present and future. *Wood Sci. Technol.*, 11, 169–218, 1977.
21. Rinaldi, R., Jastrzebski, R., Clough, M.T., Ralph, J., Kennema, M., Bruijnincx, P.C.A., Weckhuysen, B.M., Paving the way for lignin valorisation: Recent advances in bioengineering, biorefining and catalysis. *Angew. Chem. Int. Ed.*, 55, 8164–8215, 2016.
22. Argyropoulos, D.S. and Menachem, S.B., Lignin, in: *Biopolymers from renewable resources*, D.L. Kaplan (Ed.), pp. 292–322, Springer, Heidelberg, 1998.
23. Boerjan, W., Ralph, J., Baucher, M., Lignin biosynthesis. *Annu. Rev. Plant Biol.*, 54, 519–546, 2003.
24. Shuai, L., Talebi Amiri, M., Luterbacher, J.S., The influence of interunit carbon–carbon linkages during lignin upgrading. *Curr. Opin. Green Sustain. Chem.*, 2, 59–63, 2016.
25. Biermann, C.J., *Handbook of pulping and papermaking*, Academic Press, An Imprint of Elsevier, San Diego, 1996.
26. Kubo, S. and Kadla, J.F., Lignin-based carbon fibers: Effect of synthetic polymer blending on fiber properties. *J. Polym. Environ.*, 13, 97–105, 2005.
27. Tu, W.-C. and Hallett, J.P., Recent advances in the pretreatment of lignocellulosic biomass. *Curr. Opin. Green Sustain. Chem.*, 20, 11–17, 2019.
28. Laurichesse, S. and Avérous, L., Chemical modification of lignins: Towards biobased polymers. *Prog. Polym. Sci.*, 39, 1266–1290, 2014.
29. Kai, D., Tan, M.J., Chee, P.L., Chua, Y.K., Yap, Y.L., Loh, X.J., Towards lignin-based functional materials in a sustainable world. *Green Chem.*, 18, 1175–1200, 2016.
30. Vishtal, A. and Kraslawski, A., Challenges in industrial applications of technical lignins. *Bioresources*, 6, 3547–3568, 2011.
31. Shuai, L. and Saha, B., Towards high-yield lignin monomer production. *Green Chem.*, 19, 3752–3758, 2017.
32. Pizzi, A., Cameron, F.-A., van der Klashorst, G.H., Soda bagasse lignin adhesives for particleboard, in: *Adhesives from renewable resources, ACS Symposium Series*, vol. 385, Chapter 7, pp. 82–95, Washington, 1989.
33. Chiaramonti, D., Prussi, M., Ferrero, S., Oriani, L., Ottonello, P., Torre, P., Cherchi, F., Review of pretreatment processes for lignocellulosic ethanol production, and development of an innovative method. *Biomass Bioenergy*, 46, 25–35, 2012.
34. Jin, Y., Cheng, X., Zheng, Z., Preparation and characterization of phenol–formaldehyde adhesives modified with enzymatic hydrolysis lignin. *Bioresour. Technol.*, 101, 2046–2048, 2010.

35. Yang, S., Zhang, Y., Yuan, T.-Q., Sun, R.-C., Lignin-phenol-formaldehyde resin adhesives prepared with biorefinery technical lignins. *J. Appl. Polym. Sci.*, 132, 42493, 2015.
36. Zhang, W., Ma, Y., Wang, C., Li, S., Zhang, M., Chu, F., Preparation and properties of lignin–phenol–formaldehyde resins based on different biorefinery residues of agricultural biomass. *Ind. Crops Prod.*, 43, 326–333, 2013.
37. Qiao, W., Li, S., Guo, G., Han, S., Ren, S., Ma, Y., Synthesis and characterization of phenol-formaldehyde resin using enzymatic hydrolysis lignin. *J. Ind. Eng. Chem.*, 21, 1417–1422, 2015.
38. Moubarik, A., Grimi, N., Boussetta, N., Pizzi, A., Isolation and characterization of lignin from Moroccan sugar cane bagasse: Production of lignin–phenol-formaldehyde wood adhesive. *Ind. Crops Prod.*, 45, 296–302, 2013.
39. Wang, M., Leitch, M., Xu, C., Synthesis of phenol-formaldehyde resol resins using organosolv pine lignins. *Eur. Polym. J.*, 45, 3380–3388, 2009.
40. Peng, W. and Riedl, B., The chemorheology of phenol-formaldehyde thermoset resin and mixtures of the resin with lignin fillers. *Polymer*, 35, 1280–1286, 1994.
41. Xu, C. and Ferdosian, F., Lignin-based phenol–formaldehyde (LPF) resins/adhesives, in: *Conversion of lignin into bio-based chemicals and materials*, C. Xu and F. Ferdosian (Eds.), pp. 91–109, Springer Heidelberg, 2017.
42. Schuerch, C., The solvent properties of liquids and their relation to the solubility, swelling, isolation and fractionation of lignin. *J. Am. Chem. Soc.*, 74, 5061–5067, 1952.
43. Thring, R.W., Vanderlaan, M.N., Griffin, S.L., Fractionation of Alcell® lignin By sequential solvent extraction. *J. Wood Chem. Technol.*, 16, 139–154, 1996.
44. Jaaskelainen, A.S., Willberg-Keyrilainen, P., Liitia, T., Tamminen, T., Carbohydrate-free and highly soluble softwood kraft lignin fractions by aqueous acetone evaporation fractionation. *Nord. Pulp Pap. Res. J.*, 32, 485–492, 2017.
45. Vanderlaan, M.N. and Thring, R.W., Polyurethanes from Alcell® lignin fractions obtained by sequential solvent extraction. *Biomass Bioenergy*, 14, 525–531, 1998.
46. Sun, R., Tomkinson, J., Lloyd Jones, G., Fractional characterization of ash-AQ lignin by successive extraction with organic solvents from oil palm EFB fibre. *Polym. Degrad. Stab.*, 68, 111–119, 2000.
47. Li, M.-F., Sun, S.-N., Xu, F., Sun, R.-C., Sequential solvent fractionation of heterogeneous bamboo organosolv lignin for value-added application. *Sep. Purif. Technol.*, 101, 18–25, 2012.
48. An, L.L., Wang, G.H., Jia, H.Y., Liu, C.Y., Sui, W.J., Si, C.L., Fractionation of enzymatic hydrolysis lignin by sequential extraction for enhancing antioxidant performance. *Int. J. Biol. Macromol.*, 99, 674–681, 2017.

49. Jaaskelainen, A.S., Liitia, T., Mikkelson, A., Tamminen, T., Aqueous organic solvent fractionation as means to improve lignin homogeneity and purity. *Ind. Crops Prod.*, 103, 51–58, 2017.
50. Jiang, X., Savithri, D., Du, X.Y., Pawar, S., Jameel, H., Chang, H.M., Zhou, X.F., Fractionation and characterization of kraft lignin by sequential precipitation with various organic solvents. *ACS Sustain. Chem. Eng.*, 5, 835–842, 2017.
51. Kim, J.Y., Park, S.Y., Lee, J.H., Choi, I.G., Choi, J.W., Sequential solvent fractionation of lignin for selective production of monoaromatics by Ru catalyzed ethanolysis. *RSC Adv.*, 7, 53117–53125, 2017.
52. Sadeghifar, H., Wells, T., Le, R.K., Sadeghifar, F., Yuan, J.S., Ragauskas, A.J., Fractionation of organosolv lignin using acetone: Water and properties of the obtained fractions. *ACS Sustain. Chem. Eng.*, 5, 580–587, 2017.
53. Dominguez-Robles, J., Tamminen, T., Liitia, T., Peresin, M.S., Rodriguez, A., Jaaskelainen, A.S., Aqueous acetone fractionation of kraft, organosolv and soda lignins. *Int. J. Biol. Macromol.*, 106, 979–987, 2018.
54. Park, S.Y., Kim, J.Y., Youn, H.J., Choi, J.W., Fractionation of lignin macromolecules by sequential organic solvents systems and their characterization for further valuable applications. *Int. J. Biol. Macromol.*, 106, 793–802, 2018.
55. Wang, G.H., Xia, Y., Liang, B.K., Sui, W.J., Si, C.L., Successive ethanol-water fractionation of enzymatic hydrolysis lignin to concentrate its antimicrobial activity. *J. Chem. Technol. Biotechnol.*, 93, 2977–2987, 2018.
56. Tagami, A., Gioia, C., Lauberts, M., Budnyak, T., Moriana, R., Lindstrom, M.E., Sevastyanova, O., Solvent fractionation of softwood and hardwood kraft lignins for more efficient uses: Compositional, structural, thermal, antioxidant and adsorption properties. *Ind. Crops Prod.*, 129, 123–134, 2019.
57. Passoni, V., Scarica, C., Levi, M., Turri, S., Griffini, G., Fractionation of industrial softwood kraft lignin: Solvent selection as a tool for tailored material properties. *ACS Sustain. Chem. Eng.*, 4, 2232–2242, 2016.
58. Wang, Y.Y., Li, M., Wyman, C.E., Cai, C.M., Ragauskas, A.J., Fast fractionation of technical lignins by organic cosolvents. *ACS Sustain. Chem. Eng.*, 6, 6064–6072, 2018.
59. Meng, X.Z., Parikh, A., Seemala, B., Kumar, R., Pu, Y.Q., Wyman, C.E., Cai, C.M., Ragauskas, A.J., Characterization of fractional cuts of co-solvent enhanced lignocellulosic fractionation lignin isolated by sequential precipitation. *Bioresour. Technol.*, 272, 202–208, 2019.
60. Lauberts, M., Sevastyanova, O., Ponomarenko, J., Dizhbite, T., Dobele, G., Volperts, A., Lauberte, L., Telysheva, G., Fractionation of technical lignin with ionic liquids as a method for improving purity and antioxidant activity. *Ind. Crops Prod.*, 95, 512–520, 2017.
61. Allegretti, C., Fontanay, S., Krauke, Y., Luebbert, M., Strini, A., Troquet, J., Turri, S., Griffini, G., D'Arrigo, P., Fractionation of soda pulp lignin in aqueous solvent through membrane-assisted ultrafiltration. *ACS Sustain. Chem. Eng.*, 6, 9056–9064, 2018.

62. Costa, C.A.E., Pinto, P.C.R., Rodrigues, A.E., Lignin fractionation from E. Globulus kraft liquor by ultrafiltration in a three stage membrane sequence. *Sep. Purif. Technol.*, 192, 140–151, 2018.
63. Allegretti, C., Fontanay, S., Rischka, K., Strini, A., Troquet, J., Turri, S., Griffini, G., D'Arrigo, P., Two-step fractionation of a model technical lignin by combined organic solvent extraction and membrane ultrafiltration. *ACS Omega*, 4, 4615–4626, 2019.
64. Rohde, V., Boringer, S., Tubke, B., Adam, C., Dahmen, N., Schmiedl, D., Fractionation of three different lignins by thermal separation techniques-A comparative study. *Glob. Change Biol. Bioenergy*, 11, 206–217, 2019.
65. Shen, K. and Calve, L.J.A.A., Improving by fractionation-ammonium-based spent sulfite liquor for waferboard binder. *Adhes. Age*, 23, 25–29, 1980.
66. Chen, L.H., Dou, J.Z., Ma, Q.L., Li, N., Wu, R.C., Bian, H.Y., Yelle, D.J., Vuorinen, T., Fu, S.Y., Pan, X.J., Zhu, J.Y., Rapid and near-complete dissolution of wood lignin at ≤ 80 °C by a recyclable acid hydrotrope. *Sci. Adv.*, 3, e1701735, 2017.
67. Zhao, W.W., Simmons, B., Singh, S., Ragauskas, A., Cheng, G., From lignin association to nano-/micro-particle preparation: Extracting higher value of lignin. *Green Chem.*, 18, 5693–5700, 2016.
68. Gan, D., Xing, W., Jiang, L., Fang, J., Zhao, C., Ren, F., Fang, L., Wang, K., Lu, X., Plant-inspired adhesive and tough hydrogel based on Ag-Lignin nanoparticles-triggered dynamic redox catechol chemistry. *Nat. Commun.*, 10, 1487, 2019.
69. Xiao, Z., Li, Y., Wu, X., Qi, G., Li, N., Zhang, K., Wang, D., Sun, X.S., Utilization of sorghum lignin to improve adhesion strength of soy protein adhesives on wood veneer. *Ind. Crops Prod.*, 50, 501–509, 2013.
70. Argyropoulos, D.S., Quantitative Phosphorus-31 NMR Analysis of Six Soluble Lignins. *J. Wood Chem. Technol.*, 14, 65–82, 1994.
71. Hu, L., Pan, H., Zhou, Y., Zhang, M., Methods to improve lignin's reactivity as a phenol substitute and as replacement for other phenolic compounds: A brief review. *BioResources*, 6, 3515–3525, 2011.
72. An, X., Schroeder, H.A., Thompson, G.E., Demethylated kraft lignin as a substitute for phenol in wood adhesive. *Chem. Ind. For. Prod.*, 3, 36–42, 1995.
73. Wu, S. and Zhan, H., Characteristics of demethylated wheat straw soda lignin and its utilization in lignin-based phenolic formaldehyde resins. *Cellul. Chem. Technol.*, 35, 253–262, 2001.
74. Liu, Y. and Li, K., Preparation and characterization of demethylated lignin-polyethylenimine adhesives. *J. Adhes.*, 82, 593–605, 2006.
75. Li, J.J., Wang, W., Zhang, S.F., Gao, Q., Zhang, W., Li, J.Z., Preparation and characterization of lignin demethylated at atmospheric pressure and its application in fast curing biobased phenolic resins. *RSC Adv.*, 6, 67435–67443, 2016.

76. Li, J.J., Zhang, J.Z., Zhang, S.F., Gao, Q., Li, J.Z., Zhang, W., Fast curing bio-based phenolic resins via lignin demethylated under mild reaction condition. *Polymers*, 9, 428, 2017.
77. Ferhan, M. and N.Y.a.M.S., A New Method for Demethylation of Lignin from Woody Biomass using Biophysical Methods. *J. Chem. Eng. Process Tech.*, 4, 160, 2013.
78. Song, Y., Wang, Z.X., Yan, N., Zhang, R., Li, J.C., Demethylation of wheat straw alkali lignin for application in phenol formaldehyde adhesives. *Polymers*, 8, 209, 2016.
79. Wang, F., Kuai, J.T., Pan, H.S., Wang, N.N., Zhu, X.B., Study on the demethylation of enzymatic hydrolysis lignin and the properties of lignin-epoxy resin blends. *Wood Sci. Technol.*, 52, 1343–1357, 2018.
80. Sawamura, K., Tobimatsu, Y., Kamitakahara, H., Takano, T., Lignin functionalization through chemical demethylation: Preparation and tannin-Like properties of demethylated guaiacyl-type synthetic lignins. *ACS Sustain. Chem. Eng.*, 5, 5424–5431, 2017.
81. Effendi, A., Gerhauser, H., Bridgwater, A.V., Production of renewable phenolic resins by thermochemical conversion of biomass: A review. *Renew. Sust. Energ. Rev.*, 12, 2092–2116, 2008.
82. Shuai, L., Amiri, M.T., Questell-Santiago, Y.M., Héroguel, F., Li, Y., Kim, H., Meilan, R., Chapple, C., Ralph, J., Luterbacher, J.S., Formaldehyde stabilization facilitates lignin monomer production during biomass depolymerization. *Science*, 354, 329, 2016.
83. Funaoka, M., Matsubara, M., Seki, N., Fukatsu, S., Conversion of native lignin to a highly phenolic functional polymer and its separation from lignocellulosics. *Biotechnol. Bioeng.*, 46, 545–52, 1995.
84. Lin, L., Nakagame, S., Yao, Y., Shiraishi, M.Y., Shiraishi, N., Liquefaction mechanism of β-O-4 lignin model compound in the presence of phenol under acid catalysis. Part 2. Reaction behaviour and pathways. *Holzforschung*, 55, 625, 2001.
85. Wang, G., Liu, X., Zhang, J., Sui, W., Jang, J., Si, C., One-pot lignin depolymerization and activation by solid acid catalytic phenolation for lightweight phenolic foam preparation. *Ind. Crops Prod.*, 124, 216–225, 2018.
86. Jiang, X., Liu, J., Du, X., Hu, Z., Chang, H.-m., Jameel, H., Phenolation to improve lignin reactivity toward thermosets application. *ACS Sustain. Chem. Eng*., 6, 5504–5512, 2018.
87. Zhang, F., Jiang, X., Lin, J., Zhao, G., Chang, H.-m., Jameel, H., Reactivity improvement by phenolation of wheat straw lignin isolated from a biorefinery process. *New J. Chem.*, 43, 2238–2246, 2019.
88. Ma, Y., Zhao, X., Chen, X., Wang, Z., An approach to improve the application of acid-insoluble lignin from rice hull in phenol-formaldehyde resin. *Colloids Surf., A*, 377, 284–289, 2011.

89. Yang, L., Wang, X., Cui, Y., Tian, Y., Chen, H., Wang, Z., Modification of renewable resources-lignin-by three chemical methods and its applications to polyurethane foams. *Polym. Adv. Technol.*, 25, 1089–1098, 2014.
90. Lee, S.J., Kim, H.J., Cho, E.J., Song, Y., Bae, H.-J., Isolation and characterization of lignin from the oak wood bioethanol production residue for adhesives. *Int. J. Biol. Macromol.*, 72, 1056–1062, 2015.
91. Podschun, J., Saake, B., Lehnen, R., Reactivity enhancement of organosolv lignin by phenolation for improved bio-based thermosets. *Eur. Polym. J.*, 67, 1–11, 2015.
92. Podschun, J., Stuecker, A., Saake, B., Lehnen, R., Structure-function relationships in the phenolation of lignins from different sources. *ACS Sustain. Chem. Eng.*, 3, 2526–2532, 2015.
93. Podschun, J., Stuecker, A., Buchholz, R.I., Heitmann, M., Schreiber, A., Saake, B., Lehnen, R., Phenolated lignins as reactive precursors in wood veneer and particleboard adhesion. *Ind. Eng. Chem. Res.*, 55, 5231–5237, 2016.
94. Ghorbani, M., Konnerth, J., van Herwijnen, H.W.G., Zinovyev, G., Budjav, E., Silva, A.R., Liebner, F., Commercial lignosulfonates from different sulfite processes as partial phenol replacement in PF resole resins. *J. Appl. Polym. Sci.*, 135, 45893, 2018.
95. Yang, S., Wen, J.-L., Yuan, T.-Q., Sun, R.-C., Characterization and phenolation of biorefinery technical lignins for lignin-phenol-formaldehyde resin adhesive synthesis. *RSC Adv.*, 4, 57996–58004, 2014.
96. Liu, J., Zhang, Z., Chen, R., Xu, Y., Wang, C., Chu, F., Use of modified lignocellulosic butanol residue in phenol-resorcinol-formaldehyde polymers. *Bioresources*, 12, 1015–1030, 2017.
97. Zhao, M., Jing, J., Zhu, Y., Yang, X., Wang, X., Wang, Z., Preparation and performance of lignin–phenol–formaldehyde adhesives. *Int. J. Adhes. Adhes.*, 64, 163–167, 2016.
98. Younesi-Kordkheili, H., Pizzi, A., Niyatzade, G., Reduction of formaldehyde emission from particleboard by phenolated kraft lignin. *J. Adhes.*, 92, 485–497, 2016.
99. Zhang, Y., Yuan, Z., Mahmood, N., Huang, S., Xu, C., Sustainable bio-phenol-hydroxymethylfurfural resins using phenolated de-polymerized hydrolysis lignin and their application in bio-composites. *Ind. Crops Prod.*, 79, 84–90, 2016.
100. Zhao, S. and Abu-Omar, M.M., Synthesis of renewable thermoset polymers through successive lignin modification using lignin-derived phenols. *ACS Sustain. Chem. Eng.*, 5, 5059–5066, 2017.
101. Zhao, S., Huang, X., Whelton, A.J., Abu-Omar, M.M., Formaldehyde-free method for incorporating lignin into epoxy thermosets. *ACS Sustain. Chem. Eng.*, 6, 10628–10636, 2018.
102. Zhang, Y., Pang, H., Wei, D., Li, J., Li, S., Lin, X., Wang, F., Liao, B., Preparation and characterization of chemical grouting derived from lignin epoxy resin. *Eur. Polym. J.*, 118, 290–305, 2019.

103. Benar, P., Gonçalves, A.R., Mandelli, D., Schuchardt, U., Eucalyptus organosolv lignins: Study of the hydroxymethylation and use in resols. *Bioresour. Technol.*, 68, 11–16, 1999.
104. Gonçalves, A.R. and Benar, P., Hydroxymethylation and oxidation of Organosolv lignins and utilization of the products. *Bioresour. Technol.*, 79, 103–111, 2001.
105. Alonso, M.V., Oliet, M., Rodríguez, F., Astarloa, G., Echeverría, J.M., Use of a methylolated softwood ammonium lignosulfonate as partial substitute of phenol in resol resins manufacture. *J. Appl. Polym. Sci.*, 94, 643–650, 2004.
106. Ren, H. and Zhu, L.L., Comparison of the structural characterization of lignophenols and alkaline lignins before and after methylolation. *Wood Sci. Technol.*, 52, 1133–1151, 2018.
107. Zhao, L.-w., Griggs, B.F., Chen, C.-L., Gratzl, J.S., Hse, C.-Y., Utilization of softwood kraft lignin as adhesive for the manufacture of reconstituted wood. *J. Wood Chem. Technol.*, 14, 127–145, 1994.
108. Vázquez, G., González, J., Freire, S., Antorrena, G., Effect of chemical modification of lignin on the gluebond performance of lignin-phenolic resins. *Bioresour. Technol.*, 60, 191–198, 1997.
109. Cheng, S., Yuan, Z., Anderson, M., Leitch, M., Xu, C., Synthesis of biobased phenolic resins/adhesives with methylolated wood-derived bio-oil. *J. Appl. Polym. Sci.*, 126, E431–E441, 2012.
110. Wang, G.H. and Chen, H.Z., Carbohydrate elimination of alkaline-extracted lignin liquor by steam explosion and its methylolation for substitution of phenolic adhesive. *Ind. Crops Prod.*, 53, 93–101, 2014.
111. Delebecq, E., Pascault, J.-P., Boutevin, B., Ganachaud, F., On the versatility of urethane/Urea bonds: Reversibility, blocked isocyanate, and non-isocyanate polyurethane. *Chem. Rev.*, 113, 80–118, 2013.
112. Bonini, C., D'Auria, M., Emanuele, L., Ferri, R., Pucciariello, R., Sabia, A.R., Polyurethanes and polyesters from lignin. *J. Appl. Polym. Sci.*, 98, 1451–1456, 2005.
113. Hsu, O.H. and Glasser, W.G., Polyurethane adhesives and coatings from modified lignin. *Wood Sci.*, 9, 97–103, 1976.
114. Newman, W.H. and Glasser, W.G., Engineering plastics from lignin-XII. Synthesis and performance of lignin adhesives with isocyanate and melamine. *Holzforschung*, 39, 345–353, 1985.
115. Mu, W., Ben, H., Ragauskas, A., Deng, Y., Lignin pyrolysis components and upgrading—Technology review. *Bioenergy Res.*, 6, 1183–1204, 2013.
116. Shafizadeh, F. and Chin, P.P.S., Thermal deterioration of wood, in: *Wood technology: Chemical aspects*, ACS Symposium Series, vol. 43, Chapter 5, pp. 57–81, Washington, 1977.
117. Patwardhan, P.R., Brown, R.C., Shanks, B.H., Understanding the fast pyrolysis of lignin. *ChemSusChem*, 4, 1629–1636, 2011.
118. El Mansouri, N.E., Farriol, X., Salvado, J., Structural modification and characterization of lignosulfonate by a reaction in an alkaline medium for

its incorporation into phenolic resins. *J. Appl. Polym. Sci.*, 102, 3286–3292, 2006.

119. Yuan, Z.S., Cheng, S.N., Leitch, M., Xu, C.B., Hydrolytic degradation of alkaline lignin in hot-compressed water and ethanol. *Bioresour. Technol.*, 101, 9308–9313, 2010.
120. Dessbesell, L., Xu, C.B., Pulkki, R., Leitch, M., Mahmood, N., Forest biomass supply chain optimization for a biorefinery aiming to produce high-value bio-based materials and chemicals from lignin and forestry residues: A review of literature. *Can. J. For. Res.*, 47, 277–288, 2017.
121. Ragauskas, A.J., Beckham, G.T., Biddy, M.J., Chandra, R., Chen, F., Davis, M.F., Davison, B.H., Dixon, R.A., Gilna, P., Keller, M., Langan, P., Naskar, A.K., Saddler, J.N., Tschaplinski, T.J., Tuskan, G.A., Wyman, C.E., Lignin valorization: Improving lignin processing in the biorefinery. *Science*, 344, 1246843, 2014.
122. Xu, C.P., Arancon, R.A.D., Labidi, J., Luque, R., Lignin depolymerisation strategies: Towards valuable chemicals and fuels. *Chem. Soc. Rev.*, 43, 7485–7500, 2014.
123. Upton, B.M. and Kasko, A.M., Strategies for the conversion of lignin to high-value polymeric materials: Review and perspective. *Chem. Rev.*, 116, 2275–2306, 2016.
124. Liu, G., Qiu, X., Xing, D., Yang, D., Phenolation modification of wheat straw soda lignin and its utilization in lignin-based phenolic formaldehyde resins. *J. Chem. Eng. Chin. Univ.*, 21, 678–684, 2006. (in Chinese).
125. Zakzeski, J., Bruijnincx, P.C.A., Jongerius, A.L., Weckhuysen, B.M., The catalytic valorization of lignin for the production of renewable chemicals. *Chem. Rev.*, 110, 3552–3599, 2010.
126. Li, C.Z., Zhao, X.C., Wang, A.Q., Huber, G.W., Zhang, T., Catalytic transformation of lignin for the production of chemicals and fuels. *Chem. Rev.*, 115, 11559–11624, 2015.
127. Shuai, L., Sitison, J., Sadula, S., Ding, J., Thies, M.C., Saha, B., Selective C–C bond cleavage of methylene-linked lignin models and kraft lignin. *ACS Catal.*, 8, 6507–6512, 2018.

3

Green Adhesive for Industrial Applications

Priyanka E. Bhaskaran[1*], Thangavel Subramaniam[1], Gobinath Velu Kaliyannan[2], Sathish Kumar Palaniappan[3] and Rajasekar Rathanasamy[2]

[1]*Department of Mechatronics Engineering, Kongu Engineering College, Erode, India*
[2]*Department of Mechanical Engineering, Kongu Engineering College, Erode, India*
[3]*Department of Mining Engineering, Indian Institute of Technology Kharagpur, Kharagpur, India*

Abstract
In the emerging scientific world, many challenges emerge in the industrial sectors to meet the demands possessed by upgraded technologies. Especially in the research on material and composite field, analyzing and characterizing the properties using existing techniques is till marching towards to fit with exact industrial prerequisites. Due to the concern with environment and energy conservation, most of the bio waste, natural fibers, and natural polymers are considered to serve as green adhesive with enhanced mechanical properties of bonding. This chapter clearly enumerates in detail about the green adhesive categories and their specific properties with respect to their recent desired industrial applications.

Keywords: Green adhesive, bio-waste, natural fibers, polymers

3.1 Introduction

Creatures, for example, geckos, flies, arachnids, and creepy crawlies have the capacity to quickly append and segregate from fluctuated surfaces under exceptionally assorted conditions [1]. This one of a kind capacity has earned the consideration of researchers and through present day imaging procedures, point-by-point assessment of their connection cushions was completed and

**Corresponding author*: priyankabhaskaran1993@gmail.com

Inamuddin, Rajender Boddula, Mohd Imran Ahamed and Abdullah M. Asiri (eds.) Green Adhesives: Preparation, Properties and Applications, (57–84) © 2020 Scrivener Publishing LLC

uncovered progressive link structure. The regularly rising natural concerns in regard to economically built wood items make the administration of fasteners and glue a significant subject. The perpetual, oil-based nature of most engineered cements accompanies concerns with respect to unpredictable natural mixes potential wellbeing perils and non-degradability [2]. The change of sustainable biomasses into nontoxic and eco-accommodating pitches is significant for dodging the potential worries that go with oil determined under the stage of green adhesives. Owing to numerous points of interest, for example, minimal effort, high usefulness, and magnificent process ability, the modern scale use of fabrication and testing has a few constraints, including the requirement for physical or potentially synthetic alterations of protein vitality utilization, a surprising expense relieving process, poor water-obstruction, and deficient bond exhibitions. The advancement of green composites-based adhesive incorporating different scales of characteristic filaments as high-effectiveness strengthening components is ordinarily sought after since it offers a few preferences [3].

To start with, green adhesive is the most plenteous, supportable, and degradable fiber material. Second, the physiological elements of green adhesive manage the cost of plants mechanical quality, higher explicit surface zone, and predominant natural bio-inspired adhesive crossing over impacts, which speak to a portion of the alluring properties. Upgrading the mechanical property has been practiced by consolidating different green adhesive in various adhesion networks. Their poor interfacial cooperation and synthetic resin adhesive dispersity for filtering is the essential weakness to their across the board use [4]. In this regard, traditional surface-functionalization systems are utilized to tailor the fiber surface to improve the similarity with the bio-inspired network. Its inherent warm protection makes it hard to shape proficient thermally conductive pathways in the green adhesives during the pitch thermo-restoring. They require extra squeezing occasions or higher squeezing temperatures on the grounds that of their moderate restoring rate, obstructing their appropriation, as an elective answer for an eco-friendly green adhesive. In this way, other than the improvement of filaments grid connections, the warmth move effectiveness unavoidably turns into a significant factor inside the successful bond execution of the enhanced green adhesives [5].

This chapter examines structure observed to be influential attributes artificially created glue configurations; at present, accessible manufacture techniques for delivering these cement structures, just as the different testing strategies that have been utilized tentatively to portray grip execution. In the last area, current patterns and difficulties viewpoint of the green adhesives are talked about [6]. A survey intending to condense the tremendous logical learning collected for intriguing perused for green adhesive researchers, hypothetical physicists, and every single other analyst inspired by the field of bio-inspired bonding particles [7].

Fixing industry guidelines also, consistent help for green items have supported bio-waste usage for green adhesive creation. Waste materials are presently taken as fuel for little scale burning motors, while some are channeled for the creation of inexhaustible and even biodegradable things. Microbial anaerobic assimilation is currently being utilized in changing over bio-waste furthermore, inexhaustible assets into an assortment of crude materials, to be utilized for creation of polymers, sealants, and cements. Acidogenic microorganisms convert natural polymers into different synthetic concoctions or acids, including sugars, natural acids, alkali, and carbon dioxide. Helpful waste materials, for example, poly amino acetic rosin can likewise be gotten from the procedure [8]. Items, for example, hot-dissolve glues from polylactic acid (PLA) are resultant from shoddy also, promptly accessible materials for example, corn, rice, sugar beets, wheat, sugar stick, sweet potatoes, and natural issue from bioreactor wastes. In contrast to PLA, rosin is gotten from pine trees conifers for the generation of glues, inks, papers, waxes cleansers. Refining and the kraft procedure are the two principle strategies for secluding rosin. An enormous test in utilizing green adhesive lies on adjusting holding quality with practical transfer [9]. Other potential competitors incorporate poly capro lactone (PCL) as main substitution or else phenethylamine (PEA) regard to sucrose acetic acid derivation iso butyrate (SAIB).

3.2 Advanced Green Adhesives Categories—Industrial Applications

3.2.1 Keta Spire Poly Etherether Ketone Powder Coating

Solvay Specialty Polymers have propelled the new super ultra-fine structured powder evaluations of Keta Spire poly etherether ketone for waterborne coatings, powder coatings, and sap preimpregnation of constant fiber composites. The adhesive grades have an apparent middle molecule size of 29 microns, while green adhesive evaluations give the best conceivable molecule size with a perceived middle molecule size of 9 microns. The new evaluations of Keta Spire PEEK powders convey enhance synthetic opposition, solidness, and quality for requesting applications in compound preparing, oil and gas, aviation, semiconductor, medicinal services, transportation, and other modern employments [10].

3.2.2 Bio-Inspired Adhesive in Robotics Field Application

The implementation of adhesive in the emerging field of robotics paved the way to move towards the soft gripper with respect to soft manufacturing with robust control. The new research is related to the development of soft gripper to handle the surface structure having rugged and soft texture without any deviation in the stiffness of handling [11]. The degree of freedom is maintained constant by adhesive property inspired by chameleon tongue to grasp the objects without disturbing the variation on the hardness of the upholding structure. It is tested to enhance the adhesive with respect to the cohesive force existing between the soft gripper and the object. The experimenting adhesive gel is placed on the gripper containing carboxylic acid and hydrogen double bond chain to lift the objects.

Figure 3.1 shows the chameleon tongue configuration of adhesive sticky muscle supported during the movement of catching the materials irrespective of object surface with better stiffness. The modern soft manipulator is designed on the basis of chameleon tongue alignment range. In the traditional gripper design, the restriction arises to handle the object due to improper adhesive force existence [12]. But in the experimention, bio inspired adhesive soft gripper exhibits better adhesive force on the holding object with very high young's modulus to maintain the rigidity where AG indicates adhesive gripper.

The designed soft gripper is tested and experimented on various object having drastic surfaces with strong bonding without misloading.

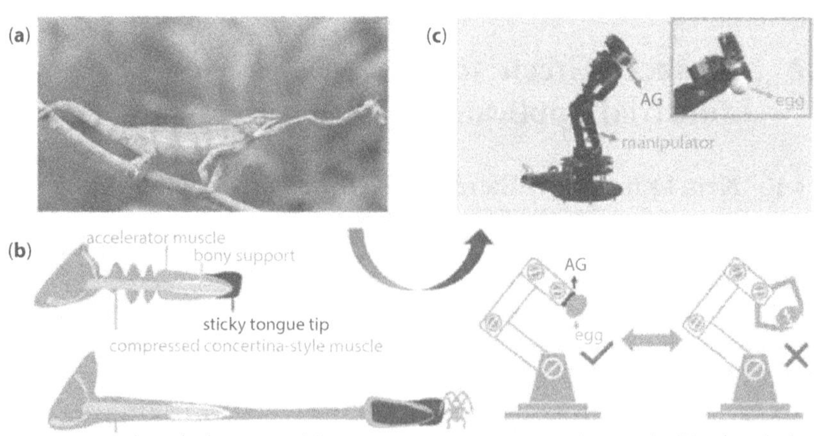

Figure 3.1 Soft gripper with bio inspired adhesive material analysis and testing [44]. (a–c) The gripper exhibits the same characterization of holding with more stiffness as like compressed accelerator modifies its style during its elongations.

The carboxylic and amino acid cross link network chain function in the adhesive glue exposes better cohesion force by expanding its cross link to the maximum extent without break in the middle hydrogen bonds [13]. The double bond structure of amino acid link carboxyl group tightly restricts the deformation when it extends the threshold limits. The experimentation is conducted up to 15 samples, but the attained results show the similar adhesive strength for all the undertaken samples. When it is exposed to the Teflon surface, the adhesive-based gripper can maintain the bonding grip up to 30 minutes. However, the experimentation is done on the wet surface for analyzing the cohesion force on the egg, it handles the egg without any deformation but its fragile nature gets increased with reduction on the bonding force [14]. Figure 3.2 clearly enumerates the entire experimentation results with final photographic view of grasping, holding, and detaching the object on the beaker by the adhesive oriented gripper obtained in the simulation platform. This bio-inspired adhesive is on the developing research stage of robotics field; further, it needs real-time analysis to compensate the carrying load and stay-tuned in the modern world circumstances [15].

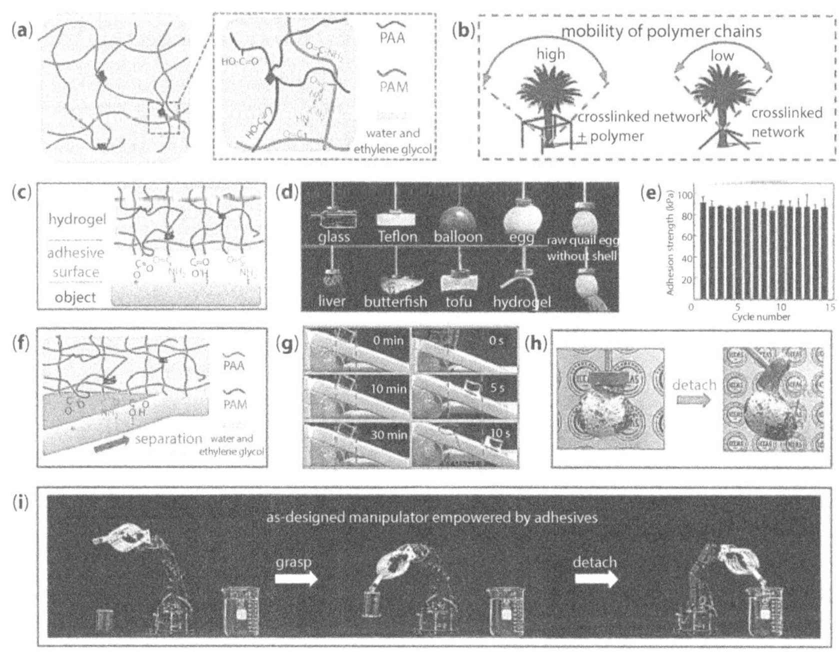

Figure 3.2 Various experimentation results of bio-inspired adhesive in soft gripper in robotics field [44]. (a–i) It shows soft gripper cohesion force capability while manipulating egg on the wet surface during grasp and detach to exemplify the nature of bio-inspired adhesives.

3.2.3 Bio-Inspired Synthetic Adhesive in Space Application

Though bio-inspired adhesive plays a major challenging role in macro-scale, synthetic dry adhesives are emerging as better solution when it is analyzed in micro and nano-scale. Fibers direction regarding the surface (tilt point) is a significant yet less examined plan on manufacture of green adhesives. Analysis including lop-sided filaments of in height viewpoint proportion showed shear attachment in anisotropic manner, intriguing component to the manufactured dry adhesives abstaining from stripping off when exposed to shear pressure [16]. Table 3.1 abridging the kinds of material utilized, young's modulus, manufacture techniques, structure, measurements, grip testing strategies, attachment quality of distributed writing from 2008 to 2019 is given in the holded application analysis. Fiber pressing balance is another significant plan thought. Experimenting geckos possess square exhibits course action to evade clustering of fibers in setae [17].

The fiber arrangement course of action give greatest bond execution because of strong bonding of filaments within grouping however could prompt serious clustering of strands when between fiber separations diminishes. Other than isotropic exhibit plans, materials with cross-link complex anisotropic behavior are observed in common connection frameworks and paved the way on increasingly refined grip orientations [18]. The preeminent case on auxiliary progression is exhibited on the gecko's foot on the milli-scale found in the lamellae, which are isolated for diverse varieties on a sufficient tiny setae, 98 μm measure on stature and 3.7 μm observation on breadth, and parting to spatulae in nano scale, 1.8 μm measure on stature and 187 nm width is given in Figure 3.3.

3.2.3.1 *Micro Structured Dry Adhesive Fabrication for Space Application*

The dry adhesive with required characterization of having fine-tuned micro or nano structure for space requirement is prepared by micro fabrication. Initially, soft molding involves pre-defined molds with natural polymers, and then, the pattern with tight panel is subjected to porosity. The substrate is wet etched with alkaline solution to remove patterned material from the soft molds. The etched substrate is subjected to chemical vapor deposition followed by photolithography processing steps [19]. The light-weight aluminium metal is put on the micro fabrication with potassium hydroxide on the etching epoxide methane substrate by photo resist membrane.

Table 3.1 Synthetic dry adhesive list with fabrication process and specific details [45].

Material	Young's modulus	Fabrication method	Structure and shape	Dimensions @ highest level (μm)	Adhesion test	Adhesion strength
PDMS	1.75 MPa	Soft molding of PDMS on SU-8 templates facricated from photolithography	Wedge shaped fibers	d = 20 or 50, h = 80 or 200	Flat contact (glass surface)	5.1 kPa (normal) 17 kPa (shear)
PDMS	2.4–3.0 MPa	Soft molding of PDMS on aluminum alloy templates	Three-tier cylindrical pillar structure	d = 300, s = 600, h = 1200	Flat contact (glass substrate)	w/ mushroom tip: ~43.6 kPA (normal) w/o mushroom tip: ~18.0 (normal)
PDMS	–	Soft molding of PDMS on Si templates followed by bonding of fiber array via dip-transfer	PDMS nanofibers on top of tilted micropillars	d = 10, s = 20, h = 5	Flat conact (Si surface)	2.35 N/cm^2 (normal) 5.96 N/cm^2 (shear)

(*Continued*)

Table 3.1 Synthetic dry adhesive list with fabrication process and specific details [45]. *(Continued)*

Material	Young's modulus	Fabrication method	Structure and shape	Dimensions @ highest level (μm)	Adhesion test	Adhesion strength
PDMS	–	Soft molding of PDMS on Si templates fabricated from photolithography and dry etching	Wide-tip micropillars	d = 5, s = 20, h = 10	Flat contact (silicone surface)	10.3 N/cm^2 (normal)
PDMS	1.5 MPa	Soft molding of PDMS on templates fabricated from photolithography followed by thermal treatment	Micropillar arrays	d = 19, s = 40, h = 13	Flat contact (Si surface)	7.09 mN (normal)
PDMS-g-poly(n-butyl acrylate) (PDMS-g-PBA)	1.51 MPa	Soft molding of PDMS on PU templates to form pillars, dip transfer to form mushroom tips then grafting of PBA chains	Mushroom-shaped pillars grafted /w PBA chains	d = 52, s = 125, h = 106	Hemispherical contact (colloidal probe-AFM, d = 6 min)	~ 14 mN (normal)

(Continued)

Table 3.1 Synthetic dry adhesive list with fabrication process and specific details [45]. (*Continued*)

Material	Young's modulus	Fabrication method	Structure and shape	Dimensions @ highest level (μm)	Adhesion test	Adhesion strength
poly(glycerol-*co*-sebacate acrylate) (PGSA)	0.38 MPa	Soft molding of PGSA on Si template fabricated from photolithography and dry etching	Conical microfibers	d = 0.18, s = 2.7, h = 2.4	Macroscale shear test	4.8 N/cm^2 (shear)
PU	3 MPa	Soft molding of PU from silicone templates fabricated by lithography followed by dip transfer	Tilted microfibers with angled mushroom tips	d = 35, s = 120, h = 100	Hemispherical contact (colloidal probe-AFM, d = 6 mm) Macroscale shear test	~5 mN (normal) ~32 mN (shear) 100 kPa (shear)
PU	–	Soft molding of PU on Si templates fabricated from photolithography and dry etching	Nanofibers with spatula tips	d = 0.41, h = 2	Flat contact (Si disk)	14.1 N/cm^2 (normal)

(*Continued*)

Table 3.1 Synthetic dry adhesive list with fabrication process and specific details [45]. (*Continued*)

Material	Young's modulus	Fabrication method	Structure and shape	Dimensions @ highest level (μm)	Adhesion test	Adhesion strength
PU	25 MPa	Soft molding of PU on PDMS templates fabricated from photolithography	Rectangular tips on top of tilted prismatic pillars	d = 20 × 7 (rectangular face), s = 37, h = 20	Macroscale shear test	1.25 N/cm^2 (normal) 2.8 N/cm^2 (shear)
PUA nanocomposites (15%Fe_3O_4 NPs /w SiO_2 shells)	20–110 MPa	Soft molding of PUA nanocomposites on Si templates	Tilted gradient pillars	d = 2.5, s = 6, h = 20	Hemispherical contact (colloidal probe-AFM, d = 4 min) Macroscale shear test	25.1 mN (shear) ~8 mN (normal) 8.8 N/cm^2 (shear)
PVS	–	Soft molding of PVS on templates followed by plasma treatment	Mushroom-shaped micropillars	d = 40, h = 70	Hemispherical contact (colloidal probe-AFM, d = 3 mm)	28.9 mN (normal)

(*Continued*)

Table 3.1 Synthetic dry adhesive list with fabrication process and specific details [45]. (*Continued*)

Material	Young's modulus	Fabrication method	Structure and shape	Dimensions @ highest level (μm)	Adhesion test	Adhesion strength
NOA81	1.36 GPa	Soft molding of NOA81 on AZ P4620 templates fabricated from photolithography	Mushroom-shaped tips on top of vertical or tilted pillars	d = 0.5, h = 1.0	Macroscale shear test	7 or 8.5 N/cm^2 (shear)
PC	2 GPa	Capillary force assisted nanoimprinting using a porous AA template	Pillars splitting into multiple fibers	d = 0.09, h = 0.85	Macroscale shear test	6.5 N/cm^2 (shear)
PS	3.2 GPa	Filling of porous AA membrane with melted PS	Pillars splitting into multiple fibers	d = 70, h = 200–700	Hemispherical contact (colloidal probe-AFM, d = 24 μm)	~1.1 μN (normal) ~8 μN (shear)
PP	1 GPa	Filling of porous PC membrane with melted PP	Microfiber array	d = 0.6, h = 20	Macroscale Shear test	2 N/cm^2 (shear)

(*Continued*)

Table 3.1 Synthetic dry adhesive list with fabrication process and specific details [45]. (*Continued*)

Material	Young's modulus	Fabrication method	Structure and shape	Dimensions @ highest level (μm)	Adhesion test	Adhesion strength
PP	1.5–2 GPa	Filling of porous PC membrane with melted PP	Tilted nanofibers on top of micropillars	d = 0.6, s = 2.77	Hemispherical contact (colloidal probe-AFM, d = 30 μm)	326 nN (normal)
High density PE (HDPE)	0.9 GPa	HDPE patterned by laser ablation and melted into porous PC membrane	Nanofiber array on top of lamellar flaps	d = 0.6, h = 18	Macroscale shear test	2.5 N/cm^2 (shear)
PE	0.4–1 GPa	Filling of porous PC membrane with melted PE	Tilted nanofibers on top of micropillars	d = 0.6	Hemispherical contact (colloidal probe-AFM, d = 30 μm)	254 nN (normal)
Pth	1.5 GPa	Electrochemical deposition of Pth on porous AA membrane	Nanotubule arrays	d = 0.2, h = 8–20	Flat contact (glass surface) Macroscale shear test	80 N/cm^2 (normal) 174 N/cm^2 (shear)

(*Continued*)

Table 3.1 Synthetic dry adhesive list with fabrication process and specific details [45]. (*Continued*)

Material	Young's modulus	Fabrication method	Structure and shape	Dimensions @ highest level (μm)	Adhesion test	Adhesion strength
Teflon	1.5 GPa	Filling of porous AA membrane with melted Teflon	Double-level Teflon nanofillars	d = 0.2, h = 37	Hemispherical contact (colloidal probe-AFM, d = 8 mm)	1.1 N/cm^2 (normal) 12 N/cm^2 (shear)
Teflon	1.5 GPa	Filling of porous AA membrane with melted Teflon	Fluffy nanostructure on top of each nanopillar	d = 0.2, h = 16	Hemispherical contact (colloidal probe-AFM, d = 6 mm)	7 mN (normal)
PMMA	–	Capillary force lithography using PUA mold	Lotus leaf-like surfaces	d = 0.15, s = 0.2, h = 0.5	Hemispherical contact (colloidal probe-AFM, d = 500 μm)	Coefficient of friction = 0.65
PDMS	4–5 GPa	Capillary force lithography	Micropillars on top of large pillars	d = 1, s = 2, h = 1	–	–

(*Continued*)

Table 3.1 Synthetic dry adhesive list with fabrication process and specific details [45]. (*Continued*)

Material	Young's modulus	Fabrication method	Structure and shape	Dimensions @ highest level (μm)	Adhesion test	Adhesion strength
CNTs	–	CVD	Vertically aligned nanofibers	d = 150, h = 0.01–0.015	Macroscale shear test	100 N/cm^2 (shear) 20 N/cm^2 (normal)
CNTs	1 TPa	CVD	Brush-like CNTs	d = ~0.008, h = 700–1000	Macroscale shear test	44.5 N/cm^2 (shear)
CNTs	4 GPa (SU-8)	VA-CNTs grown by CVD and imprinted on top of SU-8 micropillars	SU-8 vertical pillars supporting VA-CNTs	d = 0.02, h = 20	Hemispherical contact (colloidal probe-AFM, d = 1 mm)	0.5 mN (normal) 0.88 mN (shear)
Ge nanowires w/ parylene shell	100–150 GPa	Ge nanowires grown by CVD followed by parylene coating	Non-aligned nanowires	d = 0.02–0.03, h = 30	Macroscale shear test	163 N/cm^2 (shear)

(*Continued*)

Table 3.1 Synthetic dry adhesive list with fabrication process and specific details [45]. (*Continued*)

Material	Young's modulus	Fabrication method	Structure and shape	Dimensions @ highest level (μm)	Adhesion test	Adhesion strength
Nickel and photoresist	6.2 GPa	Nano/microstructures created from photolithography & dry etching, followed by plasma treatment	Nickel paddles supporting vertically aligned polymer nanorods	d = 0.2, h = 3	Flat cotact (glass punch)	14 Pa (normal)
Acrylic based negative tone resist (IPG780)	4 GPa	3D direct laser writing	Mushroom-shaped fibers on micropillars	d = 0.6, s = 1.1 or 1.4, h = 1.0 or 2.7	Hemispherical contact (colloidal probe-AFM, d = 20 μm)	~0.12 μN (normal)

A non-exhaustive list of synthetic dry adhesives produced from different materials and fabrication methods. PP – poly(ethylene), PE – poly(ethylene), PUA – poly(urethane acrylate), PDMS – Poly(dimethylsiloxane), CNT – carbon nanotube, Pth – poly(thiophene), PVS – poly(vinylsiloxane), PMMA – poly(methyl methacrylate), PC – poly(carbonate), PS – poly(styrene), AA – anodic aluminium, AFM – atomic force microscope, d – diameter, s – spacing, h – height.

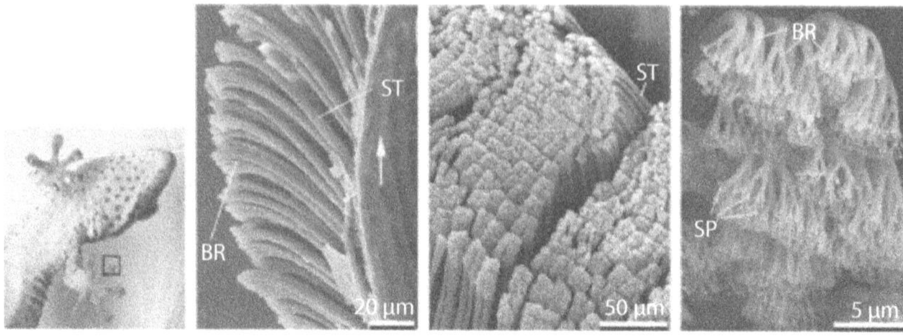

Figure 3.3 Geckos-Fiber clustering arrangement to make adhesion on the surface [46].

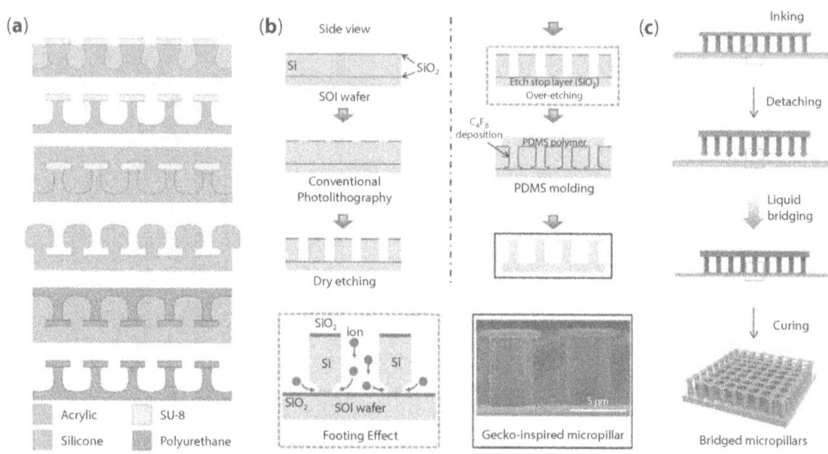

Figure 3.4 Micro fabrication process utilized for making dry adhesive in the space application [46].

A detailed micro fabrication schematic representation is given in Figure 3.4. The final manufactured synthetic adhesives have micro structured fiber as influenced by geckos' foot arrangement for better stiffness with high young's modulus to exhibit strong adhesion on the metal contacts [20]. The micro array adhesive is produced by exposing oxygen plasma to produce patterned fiber having low aspect ratio. The bio-inspired adhesive fiber arrangement as like geckos' foot adhesive holding better improved mechanical strength as compared to macro scale manufacturing related to space applications.

3.2.4 Natural Polymer Adhesive for Wood Panel Industry

Close to the issues and difficulties confronting the broad utilization of lignin-based adhesive in the wood development industry, some positive components may urge in this industry to move to lignin-based cements [21]. In expansion to the issues referenced above (maintainability and consumption of oil), another significant factor is wellbeing and natural concerns [3]. A few stringent natural and wellbeing guidelines have been concocted to deal with the earth, wellbeing, and security of purchasers. Potential wellbeing dangers have raised as of late identifying with particular kinds of crude material (synthetic compounds) normally utilized in manufactured natural polymer adhesive plans. The most usually utilized wood cements utilized in the generation of composite wood boards have been formaldehyde-based. Formaldehyde has been an imperative substance in the blend of such polymer adhesive. Figure 3.5 shows the lignin preparation by rot fungi through demethylation process [22].

Unreasonable presentation of people and the condition to cancer-causing formaldehyde is presently perceived as a serious issue. Saps in manufactured polymer adhesive are frequently broken down in natural solvents which fill in as transporters with the goal that the pitches can be connected effectively. Such natural solvents can include up to 80% of the last weight of the item and this is the fundamental issue [23]. Unpredictable natural solvents that transmit hurtful gases are terrible for air quality. In this manner, wood adhesive makers have gone under administrative strain to change the detailing of engineered oil-based adhesive substrate to follow those guidelines. Natural polymer resin-based adhesive utilized in wooden items may turn into a snag for reusing after culmination of their administration life expectancy. Resin adhesives are one of the considered contaminants other than paint and wood additives which would hinder the capacity to reuse, reuse or now and again diminish scope for vitality recuperation from the recovered wood asset [24]. Whenever designed wooden items are reinforced with bio-based adhesives, this could be an appealing way where utilized wooden items could be sent for fertilizing the soil, reuse or recuperation of vitality without settling natural, or wellbeing and security issues.

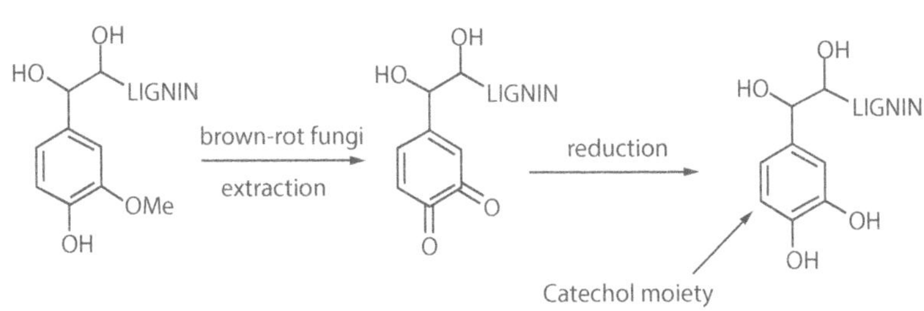

Figure 3.5 Lignin formation through rot-fungi involved technique.

A recent investigating engineered oil-based adhesives keep on being, by far, the most significant wood panel comprising adhesive utilized by makers in this industry, the variance of cost, limited and non-sustainable wellsprings of oil and open familiarity with the wellbeing, security and natural concerns are the significant components that may constrain the use of such chemical based synthetic resin adhesive [25]. Henceforth, the eventual fate of lignin-based polymer adhesive materials in this industry by the broad investigations that have been led by numerous scientists demonstrating the agreeable execution of lignin-based adhesive supporting substrate for composite wood boards. Be that as it may, a few noteworthy issues, particularly the greater expenses of utilizing lignins in wood adhesive must be survived to energize the enormous scale commercialization of lignin-based copolymer.

3.2.5 Tannin Based Bio-Adhesive for Leather Tanning Industry

Tannins as home grown concentrates are a decent elective for green science. They are found in abundance in some plant species and can be extricated by straightforward strategies. The tannins fundamental component and favorable position is their phenolic structure which is fundamentally the same as engineered phenols [26]. Another key element of tannins is their activities as cancer prevention agents. The utilization of tannins is frequently associated with one of these highlights to get commercialized in leather industry.

In spite of the fact that oil-sourced materials ended up far reaching in the previous few decades, they are at present subject to bunches of addressing due to the fluctuating oil cost, to issues of stores consumption, to the contamination which stems from these, and so on. These days, practical materials (for example, polylactic corrosive, chitosan, lignin, tannins, and so on) are increasing an expanding significance so as to decrease raw petroleum utilization and to lessen human natural effect. Among them, tannins are one of the most intriguing materials that can be acquired from plants. They are common phenolic structures which are found richly in the bark just as wood and in a lesser sum in leaves also, products of an assortment of plant animal types. Tannins job in plants incorporate security against light (UV beams and free radicals) and protection on account of various organic dangers (creatures, creepy crawlies, parasites, and microscopic organisms). They are otherwise called shielding operators of certain plants from dryness, through sub-atomic systems, as on account of the revival plants [27]. Tannins are commonly grouped into dense and hydrolysable tannins. Hydrolysable tannins are blends of basic phenols, as the name shows hydrolysable tannins can be hydrolyzed by powerless acids/bases to deliver sugar and phenolic acids. Some plant species contain either gallo-tannins or ellagi-tannins and some others contain a wide range of hydrolysable and consolidated tannins. For instance Acer, Acacia, and Quercus species have been perceived as species which contain the two sorts of tannins.

Tanning is one of the most significant stages in changing covers up to leather and as its name proposes, it is traditionally done utilizing tannins [28]. This procedure gives quality, delicateness, and obstruction against synthetic compounds, warm strength, and antimicrobial properties to calfskin. Chrome tanning is one of the basic mechanical tanning forms since it gives delicate, light, and an adaptable leather with great warm strength and without breaking a sweat of preparing which is difficult to accomplish with other assembling techniques. Then again, chromium is perceived as being conceivably cancer-causing and when it is released as waste it causes soil defilement. A portion of different issues with chrome-based tanning incorporates constrains in recuperation and reuse of dissipates from leather production and constrained accessible sources, more contamination, complexities in waste treatment, increment in request of natural oxygen just as request on artificial oxygen (COD). Figure 3.6 gives the industrial oriented preparation step for the tannin compounds [29].

One of the fundamental difficulties that this industry is associated with is the contaminations which originate from chromium utilization. Because of natural concerns and new guidelines, the leather business is turning again toward utilizing greener and cleaner materials and procedures as

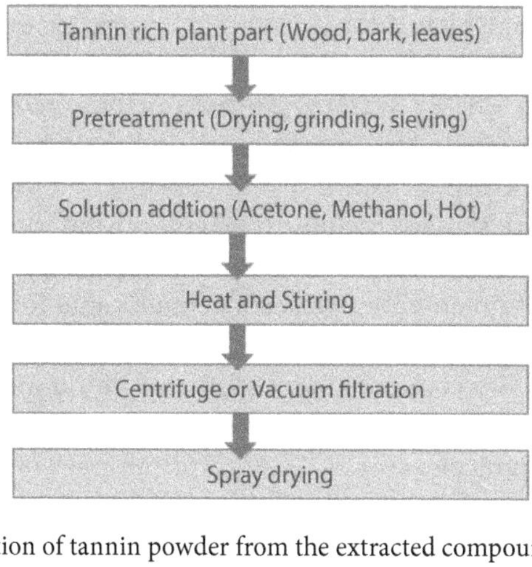

Figure 3.6 Preparation of tannin powder from the extracted compounds.

opposed to utilizing chromium based tanning forms. Vegetable tannins are great substitute for chromium and the blend of these two frameworks has been considered as a satisfactory cleaner arrangement [30].

At that point, fundamental protein in creatures is collagen, which can be found in bounty in skin. Chromium mixes are known to respond well with collagen and results in lasting changes. These progressions produce fantastic leather from biodegradable stows away. Tannin polyphenols can make crosslinking with collagen by various hydrogen bonds. Moreover, the response of tannins with collagen brings about a progressively steady leather texture. Examinations have demonstrated that increasingly warm stable leather is feasible with vegetable tannins and acrylic polymer [31].

3.2.6 Conductive Adhesives in Microelectronics Industry

Microelectronic innovations and businesses have progressed in significant degree in the course of the most recent printed circuits on multi-layer arrangement, dainty motion picture transistors, sun powered modules, and straightforward electronic covering. In any case, the intrinsic poisonous quality of existing coating composites on wellbeing formulation and condition have turned into defeat to afford powers numerous nations on force severe enactments with confinements on quell the utilization of conventional composites in electronic apparatuses [32].

Until towards the upgraded electronics industries, the bind combination of synthetic resin adhesive with bio-inspired adhesive possess string toughness on the printed board surfaces is implemented in hardware enterprises. Electrically oriented conductive nature adhesives are developed from pentagon linked poly-chain arrangement with metal fillers. The lattice plays vital contribution which affords isolation property on diverse surface existing between green adhesive. Based on the inter-connected filler connection emerging on the surface printed on the circuit board provides electrical connections to the nearby layer. The double usefulness accomplishment is by affixing dielectric resist deposition on the conductive electrical substrate with same as metallic bind composites for empowering high mechanical characterization and free mobility [33]. By and by, mechanical progression of green adhesive seems conceivable for supplanting customary patch composites with traditional industrial adhesives because of reduced conductivity contrasted and customary bind junctions and moderately lesser thermal characterization by raised surrounding temperature.

The most economically critical ACAs (Anisotropic Conductive Adhesive) depend on the single-molecule crossing over idea. Interconnections with pitches beneath 48 mm have been shown with these materials. The undertaken ACA is connected to the deposition layer, holding a variety of cohesion bond force. On completing testimony, a segment is adjusted to the deposition layer and subjected for conductive adhesive [34]. By taking weight in respect to heat is connected at the same time with the end goal that abundance bonding force is dislodged and every mobility particles are caught between the etching layer and photo resist shaping have bonded numerous conductivity interconnections. During fixing the adhesive, shrinkage with respect to compression on cooling by vigorous attachment keep up the metallization interface on the surface that is basic for dependable electric-mobility contacts. Circular elements (perspective proportion 0.82) are utilized to bring down the likelihood of electro-resistivity on the X-Y plane as shown in Figure 3.7. The utilization of tight molecule size conveyances additionally lessens the likelihood of permeation toward this path on an oriented fixation [43].

Numerous adjustments carried on essential methodology created to shape anisotropic conductive adhesive arrangement at (44 mm) well-tuned pitches [35]. More often than not, these methodologies endeavor to limit transference on the applied X-Y plane. In certain techniques, adjusting particles having metallic nature in an attractive or power-driven field upgrades anisotropic behavior and grants empowered molecule loadings.

Different techniques accumulate the metallic particles with a dainty dielectric layer with polymer. The covering cracks over heads when it is condensed between two roughing surfaces, rendering the coated surface to be electrically conductive. An important way to deal with fine scaled pitch configurations stores atoms just over contact bonded natural bio-adhesive [36].

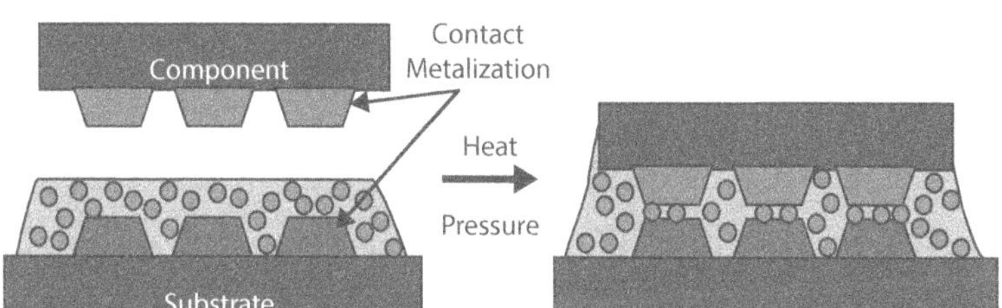

Figure 3.7 Assembly process to attain conductive adhesive by heat and pressure displacement.

3.2.7 Bio-Resin Adhesive in Dental Industry

Adhesive therapeutic dentistry influences for all intents and purposes each dental practice since it is fundamental to numerous methods, including dental sealant position, holding of orthodontic sections, direct composites reclamations, intra radicular posts adhesion, or tooth-shaded reclamations, full-inclusion all-artistic crowns, reinforced extensions, and root waterway shield bondage.

In general, attachment is achievable when coming up next are available: clean dental surfaces, great surface wettability, dispersion of the cement gum monomers inside lacquer and dentin, and sufficient pitch polymerization. The cement frameworks comprise of a mix of methacrylate-based sap monomers with either 2 (cross-connecting monomers) or 1 (practical monomers) poly merizable closures, natural solvents, a photo initiator framework, also, frequently nano fillers [37]. The science of dental cement gums must satisfy the prerequisites for bond to various dental substrates: lacquer, dentin, and cementum. Figure 3.8 shows the lowest to highest dental resin interfaces in practical oral-dental application procedures.

Utilitarian hydrophilic pitch monomers encourage gum invasion inside the demineralized and clammy dentin surface while hydrophobic cross-connecting pitch monomers give the mechanical quality, security, and similarity between the cement framework and the mass remedial pitch or tar cement. Pitch monomers with at least 2 polymerizable gatherings are important to shape an exceptionally cross-connected system to give the quality and strength of the cement layer. Examples of cross-connecting monomers with a progressively hydrophobic nature are bisphenol glycol dymethacrylate. The distinction in sub-atomic load among the tar monomers is significant, in light of the fact that low-atomic weight monomers break up the high-atomic weight monomers, improving the wettability of the gum mix [38]. Practical monomers as a rule have 1 useful gathering

Contemporary Dental Adhesive Systems					Characteristics			Longevity
System Mode	Delivery	Adhesion Steps: Etching	Primer	Adhesive	Acidity	Hydrophilicity	Bond Stability[b]	
Etch-and-rinse	3-step				+	+	++++	Stability
	2-step				++	++	+++	Degree of Conversion Solvent Evaporation
Self-etch	2-step				+++	++	++++	Acidity Hydrophilicity
	1-step				++++	+++	+	Degradation
Universal	1 or 2 steps[a]				+++	++	+(+)+	

Figure 3.8 Practical implementation of dental resin interface regarding adhesives [39].

(eg, hydroxyl gatherings) and a single polymerizable gathering to shape straight polymer chains.

3.2.8 Green Adhesive in Fiberboard Industry

Fiberboard (additionally called boards) comprises of lignocellulosic strands and normally a manufactured sap with reasonable holding framework fortified with heat and weight. On the basis of the thickness, fiberboards can be characterized in medium and high thickness fiberboards. Generally, on creating fiberboards from wood as predominant material, however utilizing an elective wellspring of materials possessing lignocellulosic, for example, agrarian deposits or then again non-wood filaments, because predominantly formed from biomass of woody sources [39]. Requirement of wood with fiber-based nature rather from prepared non-eco-friendly complex synthetic polymer compounds. In recent sense, focused reasonable necessity of creating fiber-oriented materials from agrarian buildups or natural woody-waste paper. In the ongoing process, a wide assortment of agrarian deposits such as filaments of non-wood fabricates fireboards with green adhesives.

Lignin with fibers structure with closed network of cellulose possesses various viscous force on the applied surface layer based on the friction with high stress force. The roughness and erectness are directly superimposed on the nano fibers having high cellulose content to form the fiberboard in large scale. The extraction wastes obtained from different wood products have great impact on the board fabrication utilizing bio-inspired fiber adhesive. The heat restriction on the viscous force breaks gets condensed on analyzing strong cohesion between fiber and deposition composite multi-layers [40].

Nonetheless, all things considered, CNF were most certainly not acting just as cement, yet in addition as fortifying operator. Different bio-adhesives could be soy protein by tannin comparison, starch with processing of glycerol and lignin. In all scenario, it was exhibited that most of the ecologically well-disposed petrochemical based, demonstrating main necessity of executing the utilization of such strong cohesive adhesive materials.

Lignin contributes as sweet-smelling compound having polymer natureas additionally an economical compound contrasted and other manufactured saps about as a characteristic fastener for filaments [41]. Considering all these, lignin is displayed as a legitimate option for maintaining a strategic distance from the utilization of urea formaldehyde. The objective of investigation was to deliver fiberboards on basis of wheat straw composite foundation including soft drink lignin with a characteristic compression in the board-production method. The impact of lignin (went from 0% to 14.7% on the underlying dry load of the mash) due to which polished physical and mechanical properties was additionally contemplated [34].

From Figures 3.9 and 3.10, it is inferred that due to the thickness worth increments as the lignin rate builds, it is intriguing stay away from the impacts of mechanical characterization. Likewise, the fuse of lignin comes to be recently answered to expand orientation dependability of wood

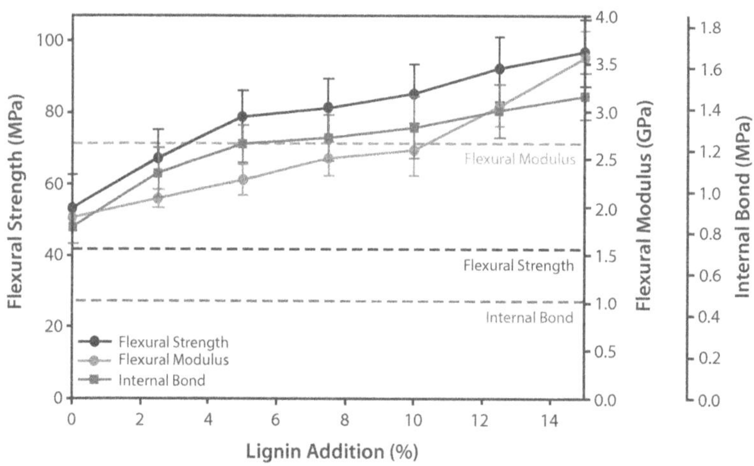

Figure 3.9 Mechanical characterization of fiberboard [34].

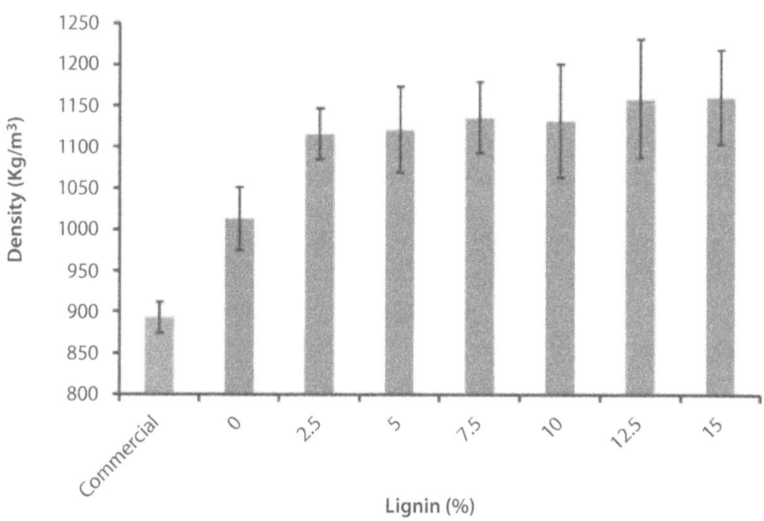

Figure 3.10 Density characterization of fiberboard [34].

dependent fiberboards in nearness of water or on the other hand stickiness, which is a critical perspective when delivering fiberboard [26].

Wheat straw serves as an astounding elective crude substance to get strands holding cellulose which is utilized for the assembling of fiberboards. The board developed uniquely by wheat straw filaments demonstrated greater flexural strength quality worth including for the business sheets. Furthermore, the utilization of such crude material for fiberboards fabricating obviously conceals wood utilization, typically considered as waste which could be submerged in the dumped field [42].

On the great concern with mechanical properties enhanced by expanding lignin rate, it very well may be expecting lignin as an encouraging option in contrast to manufactured cements on the generation of composites with lignocellulosic content. The experimented lignin test conducted in a similar procedure where cellulosic filaments, not representing fundamental of extra cleaning venture to accomplish fascinating outcomes. As final to create the examined environmentally based fiberboards advances a superior misuse on biomass holding lignocellulosic specification utilized. Furthermore, the utilization of lignin strands keeps away from auxiliary and extra forms, making the entire procedure progressively productive [19].

3.3 Conclusions and Future Scope

Critical upgrades to the science and treatment of cement frameworks have expanded the utilization of green adhesives in major industry applications. This chapter gives brief outline about the green adhesive in the major industries where the bio-adhesive serves better with its enhanced mechanical characterization by modern fabrication technology. Each bio-inspired and synthetic polymer adhesive exhibits key contributions in advanced industrial applications by its strong adhesion and cohesion force to attain high binding stiffness. Late endeavors have additionally observed the consolidation of cutting edge highlights, for example, micro and nano-molded structures especially to improve contact pressure conveyance and attachment directionality to encourage separation.

References

1. Youssef, A.M., Assem, F., Essam, M., Elaaser, M., Ibrahim, O., Mahmoud, M., Abd ElSalam, M., Development of a novel bio nanocomposite material and its use inpackaging of Ras cheese. *Food Chem.*, 270, 467–75, 2019.

2. Youssef, A.M., El-Sayed, S.M., El-Sayed, H.S., Salama, H.H., Mohamed, H.F., Abd ElSalam, M.A., Novel bio nanocomposite materials used for packaging skimmed milkacid coagulated cheese (Karish). *Int. J. Biol. Macromol.*, 115, 1002–11, 2018.
3. Youssef, A.M., El-Nahrawy, A.M., Abou Hammad, A.B., Sol-gel synthesis and characterizations of hybrid chitosan-PEG/calcium silicate nanocomposite modified with ZnO-NPs and (E102) for optical and antibacterial applications. *Int. J. Biol. Macromol.*, 97, 561–7, 2017.
4. Youssef, A.M. and El- Sayed, S.M., Bio nanocomposites materials for food packagingapplications: Concepts and future Outlook. *Carbohydr. Polym.*, 193, 19–27, 2018.
5. Bumbudsanpharoke, N. and Ko, S., Nano-food packaging: An overview of market,migration research, and safety regulations. *J. Food Sci.*, 80, R910, 2015.
6. Youssef, A.M., Polymer nanocomposites as a new trend for packaging applications. *Polym. Plast. Technol. Eng.*, 52, 635–60, 2013.
7. Yam, K.L., Takhistov, P.T., Miltz, J., Intelligent packaging: Concepts and applications. *J. Food Sci.*, 70, R1–10, 2005.
8. Venugopal, J. and Ramakrishna, S., Biocompatible nanofiber matrices for the engineering of a dermal substitute for skin regeneration. *Tissue Eng.*, 11, 847–54, 2005.
9. Moustafa, H., El Kissi, N., Abou-Kandil, A.I., Abdel-Aziz, M., Dufresne, A., PLA/PBATbio nanocomposites with antimicrobial natural rosin for green packaging. *ACS Appl. Mater. Interfaces*, 9, 20132–41, 2017.
10. Moustafa, H., Duquesne, S., Haidar, B., Valla, M.F., Influence of the degree ofexfoliation of an organoclay on the flame-retardant properties of cross-linked ethylene-co-Propylene-co-diene monomer-g-Maleic anhydride-based composites. *Polym. Compos.*, 38, 966–73, 2017.
11. Moustafa, H., Darwish, N.A., Nour, M.A., Youssef, A.M., Biodegradable date stones fillerfor enhancing mechanical, dynamic, and flame retardant properties of polyamide-6 bio composites. *Polym. Compos.*, 39, 1978–87, 2018.
12. Siqueira, G., Bras, J., Dufresne, A., Cellulosic bio nanocomposites: A review ofpreparation, properties and applications. *Polymers*, 27, 28–765, 2010.
13. Riaz, U. and Ashraf, S.M., Plant oil renewable-resource-based biodegradable blends asgreen alternatives in bio packaging. *Int. J. Polym. Mater. Polym. Biomater.*, 61, 3, 229–39, 2012.
14. Seltenrich, N., A hard nut to crack: Reducing chemical migration in food contactmaterials. *Environ. Health Perspect.*, 123, A174–9, 2015.
15. Neethirajan, S. and Jayas, D.S., Nanotechnology for the food and bioprocessingindustries. *Food Bioprocess. Technol.*, 4, 39–47, 2011.
16. Youssef, A.M., Abdel-Aziz, M.S., El-Sayed, M.S., Chitosan nanocomposite films based on Ag-NP and Au-NP biosynthesis by *Bacillus subtilis* as packaging material. *Int. J. Biol. Macromol.*, 69, 185–91, 2014.

17. Fang, Z., Zhu, H., Yuan, Y., Ha, D., Zhu, S., Preston, C., Chen, Q., Li, Y., Han, X., Lee, S., Novel nanostructured paper with ultrahigh transparency and ultrahigh haze for solar cells. *Nano Lett.*, 14, 765–73, 2014.
18. Hu, L., Zheng, G., Yao, J., Liu, N., Weil, B., Eskilsson, M., Karabulut, E., Ruan, Z., Fan, S., Bloking, J.T., McGehee, M.D., Wagberg, L., Cui, Y., Transparent and conductive paper from nano cellulose fibers. *Energy Environ. Sci.*, 6, 513–8, 2013.
19. Luo, Y., Zhang, J., Li, X., Liao, C., Li, X., The cellulose nanofibers for optoelectronic conversion and energy storage. *J. Nanomater.*, 2014, 13, 2014.
20. Youssef, A.M., Youssef, M., Ayad, D.M., Sarhan, A.A., A novel approach to prepare Poly (vinyl acetate)/Ag nanocomposite for effective antimicrobial coating applications. *Polym. Plast. Technol. Eng.*, 54, 1735–42, 2015.
21. Zhou, Y., Khan, T.M., Liu, J.-C., Fuentes-Hernandez, C., Shim, J.W., Najafabadi, E., Youngblood, J.P., Moon, R.J., Kippelen, B., Efficient recyclable organic solar cells on cellulose nanocrystal substrates with a conducting polymer top electrode deposited by flm-transfer lamination. *Org. Electron.*, 15, 661–6, 2014.
22. Belli, B., Eating nano: Processed foods and food packaging already contain nanoparticles-Some of which could be harmful to our health, The Environmental Magazine Website, Poland, 2012, Available from: http://www.emagazine.com/includes/print-article/magazine/9623/; Accessed.
23. Aguilera, J.M., Where is the nano in our foods? *J. Agric. Food Chem.*, 62, 9953–6, 2014.
24. Tuinier, R. and de Kruif, C.G., Stability of casein micelles in milk. *J. Chem. Phys.*, 117, 1290–5, 2002.
25. Zhang, L., Chen, F., An, H., Yang, H., Sun, X., Guo, X., Li, L., Physicochemical properties, firmness, and nanostructures of sodium carbonate-soluble pectin of 2 Chinese cherry cultivars at 2 ripening stages. *J. Food Sci.*, 73, N17–22, 2008.
26. Moustafa, H., Galliard, H., Vidal, L., Dufresne, A., Facile modification of organoclay and its effect on the compatibility and properties of novel biodegradable PBE/PBAT nanocomposites. *Eur. Polym. J.*, 87, 188–99, 2017.
27. Moustafa, H., Guizani, C., Dufresne, A., Sustainable biodegradable coffee grounds fller and its effect on the hydrophobicity, mechanical and thermal properties of biodegradable PBAT composites. *J. Appl. Polym. Sci.*, 134, 44498, 2017.
28. Youssef, A.M., Bujdoso, T., Hornok, V., Papp, S., Kiss, B., Abd El-Hakim, A., d ekany, I., Structural and thermal properties of polystyrene nanocomposites containing hydrophilic and hydrophobic layered double hydroxide. *Appl. Clay Sci.*, 77–78, 46–51, 2013.
29. Abd El-Ghaffar, M.A., Youssef, A.M., Abd El-Hakim, A.A., Polyaniline nanocomposites via *in-situ* emulsion polymerization based on montmorillonite; preparation & characterization. *Arab. J. Chem.*, 8, 771–9, 2015.

30. Da Silva, C., Haidar, B., Vidal, A., Miehe-brendle, J., Le Dred, R., Vidal, L., Preparation of EPDM/synthetic montmorillonite nanocomposites by direct compounding. *J. Mater. Sci. Lett.*, 40, 1813–5, 2005.
31. Wang, L., Li, X., Zhang, G., Dong, J., Eastoe, J., Oil-in-water nano emulsions for pesticide formulations. *J. Colloid Interface Sci.*, 314, 230–5, 2007.
32. Fornes, T.D., Yoon, P.J., Hunter, D.L., Keskkula, H., Paul, D.R., Effect of organoclay structure on nylon 6 nanocomposite morphology and properties. *Polymer*, 43, 5915–33, 2002.
33. Vaia, R.A. and Giannelis, E.P., Lattice model of polymer melt intercalation in organically modified layered silicates. *Macromolecules*, 30, 7990–9, 1997.
34. Domínguez-Robles, J., Tarrés, Q., Delgado-Aguilar, M., Rodríguez, A., Espinach, F.X., Mutjé, P., Approaching a new generation of fiberboards taking advantage of self lignin as green adhesive. *Int. J. Biol. Macromol.*, 108, 927–935, 2018.
35. Dufresne, A., Cellulose nanomaterial reinforced polymer nanocomposites. *Curr. Opin. Colloid Interface Sci.*, 29, 1–8, 2017.
36. Kumar, N. and Kumbhat, S., Carbon-based nanomaterials, in: *Essentials in nanoscience and nanotechnology*, pp. 189–236 chapter 5, John Wiley & Sons, Inc., Hoboken, NJ, 2016.
37. Badrossamay, M.R., McIlwee, H.A., Goss, J.A., Parker, K.K., Nanofber assembly by rotary jet-spinning. *Nano Lett.*, 10, 2257–61, 2010.
38. Gleiter, H., Nanostructured materials: Basic concepts and microstructure. *Acta Mater.*, 48, 1–29, 2000.
39. Bedran-Russo, A., Leme-Kraus, A.A., MP Vidal, C., Teixeira, E.C., An Overview of Dental Adhesive Systems and the Dynamic Tooth–Adhesive Interface. *Dent. Clin.*, 61, 4, 713–731, 2017.
40. Gokarna, A., Parize, R., Kadiri, H., Nomenyo, K., Patriarche, G., Miska, P., Lerondel, G., Highly crystalline urchin-like structures made of ultra-thin zinc oxide nanowires. *RSC Adv.*, 4, 47234–9, 2014.
41. Tiwari, J.N., Tiwari, R.N., Kim, K.S., Zero-dimensional, one-dimensional, two dimensional and three-dimensional nanostructured materials for advanced electrochemical energy devices. *Prog. Mater. Sci.*, 57, 724–803, 2012.
42. Adeosun, S.O., Lawal, G.I., Balogun, S.A., Akpan, E.I., Review of green polymer nanocomposites. *J. Miner. Mater. Char. Eng.*, 11, 385–416, 2012.
43. Anbazhagan, M., Ramachandran, S., Subramanian, P., Nachimuthu, R., Gothandam, K.M., Ranjan, S., Dasgupta, N., Lichtfouse, E., *Nanoscience in food and agriculture "nanomaterials: Classifcation, biological synthesis and characterization"*, Third ed., pp. 31–71, Springer, Atlanta, USA, 2016.
44. Wang, C., Xu, L., Qiao, Y., Qiu, D., Adhesives to empower a manipulator inspired by the chameleon tongue. *Chin. Chem. Lett.*, 234, 09, 890–903, 2019.
45. Pan, H.M., Bioinspired Adhesives, 2019.

4

Green Adhesives for Biomedical Applications

Julliana Ribeiro Alves dos Santos[1], Alessandra Teixeira de Macedo[1], Audirene Amorim Santana[2], Maria Eliziane Pires de Souza[3], Rodrigo Assuncao Holanda[1] and Glauber Cruz[3*]

[1]*Ceuma University, Environment Microbiology Laboratory, São Luís, Brazil*
[2]*Federal University of Maranhão, Department of Chemical Engineering, São Luís, Brazil*
[3]*Federal University of Maranhão, Department of Mechanical Engineering, São Luís, Brazil*

Abstract

There is a consensus among societies worldwide that new biomaterials have gained high significance over the last few decades because of their biocompatibility, bio-inertness, non-toxicity, and non-immunogenicity among other properties, which enable their varied applicability. Green adhesives produced from polysaccharides or biopolymers are renewable materials that are readily available and play an important role in the medical field for their several applications, for instance, tissue engineering, drug and gene delivery, wound healing, among others. However, for a wider, safe, and reliable application of these biomaterials, some vital properties need to be analyzed. As a result, this study revisits chemical, physical, thermal, mechanical, and biological characteristics of four green adhesives: alginate, chitosan, lignin, and PLA and their blends for biomedicine employment. Our study also presents different characterization techniques to gather precise information of the aforementioned adhesives, besides presenting the results of other studies in this field. Finally, it describes new technologies, as well as possible and future applications, for the green adhesives, in the biomedical and industrial fields.

Keywords: Biomedicine, properties, synthesis, technology, tissue engineering

**Corresponding author*: cruz.glauber@ufma.br

Inamuddin, Rajender Boddula, Mohd Imran Ahamed and Abdullah M. Asiri (eds.) Green Adhesives: Preparation, Properties, and Applications, (85–120) © 2020 Scrivener Publishing LLC

4.1 Introduction

The use of adhesives in medical procedures dated way back to 1960, when an adhesive was used for blood vessels [1]. Since 1960, the use of applications involving adhesives has been on the increase; few examples of applications are abdominal, orthopedic, and vascular surgeries that require soft tissue adhesives [1]. These adhesives are important where sutures do not hold wounds. Tissue adhesives are also used in laparoscopy and robotic surgery [2]. Adhesives and sealants instead of sutures and staples can be used to prevent air and liquid leakages during surgeries [3, 4]. Another important application of adhesives in medicines includes drug delivery systems [5–8].

In general, to use an adhesive, material characteristics and properties are analyzed for favorable results in terms of device performance. Adhesiveness is the most important property required for proper performance. Such properties may be directly related to the mechanical behaviors, such as shear and traction, of the adhesive and in contact with the region of the organism where it will act. According to Zhu *et al.* [9] and Marshall *et al.* [10], bonding occurs in a bioadhesive as a result of adhesive and cohesive forces. Adhesive forces occur between the adhesive and the surface of the tissue or region that has bonding components, whereas cohesive forces result from the internal molecular forces of the bioadhesive, which are important to withstand any external forces that may arise [10]. A better adhesive result can be obtained when there is a balance between adhesive and cohesive forces [1].

Factors such as material type, bond stability, roughness and porosity of surfaces in contact, and environment may affect the behavior of bioadhesives [11]. Owing to this, in terms of mechanical performance, properties such as tensile stress and shear modulus are somewhat allied to describe the best performance of adhesives [3, 5, 12, 13]. Besides that, the relation between body fluids and adhesives can be explored through contact angle measurements, which describe the hydrophilic behavior; this is important for optimal performance as most clinical applications are used in a moist environment [14, 15].

The last important factor is biocompatibility, which predicts the response of the body to the patch. An adhesive when inserted into a body should not cause any inflammation in the body or should not be rejected by the body [16, 17]. Owing to this, green adhesives have been studied and turned out to be an important solution, because they originate from renewable natural resources and thus have a lower tendency to be rejected by a body [5, 18].

Combinations of different materials had to be employed to biomedical applications, and the properties sought are similar, regardless of location or type of adhesive. As a result, polymeric materials have been employed because such materials can generate three-dimensional (3D) networks that physically or chemically connect to target tissues and act as hemostats, sealants, or adhesives [16]. Polymeric materials are not always safe or environmentally friendly. Different raw materials may be employed for the production of such materials which may occur by condensation polymerization, polymerization of hydrocarbons or similar hydrocarbon monomers [19]. The petrochemical origin makes these materials considered non-renewable and non-biodegradable. Over the past three decades, there had been an increasing interest in the development of biopolymers to find an alternative to the aforementioned polymers.

Biopolymers derive from living organisms or are synthesized from renewable resources [20]. Biological origin and non-toxic property of biopolymers for biomedical applications minimize the environmental impacts regarding waste handling. Unlike the degradation processes of conventional polymers, in biological polymers, the action of microorganisms occurs faster. Materials for their production include polysaccharides, proteins, and lipids [21]. In recent years, various ways of using materials made from biopolymers have been developed, resulting in the progressive development of sustainable biotechnology and bioeconomy [22].

The medical and biomedical use of polymers is essential and is constantly growing in the industrial and research sectors. Biomedical engineering uses techniques related to biology, engineering, medicine, physics, and chemistry to develop therapies and technologies that repair or replace damaged cells, tissues, and organs. Thus, a new generation of biodegradable analog and synthetic natural polymers has emerged [23].

Different natural materials have been studied and applied to produce green adhesives, also known as bioadhesives, for medical applications. Being a new subject, in this chapter, we intend to present some raw materials that have been used in the preparation of bioadhesives, as well as the most common characterization techniques. In addition, in the chapter, we will discuss significant applications of bioadhesives in medicines.

4.2 Main Raw Materials of Green Adhesives: Structure, Composition, and Properties

It is known that petrochemical industry is the main precursor of the polymeric materials and these are considered as non-renewable and

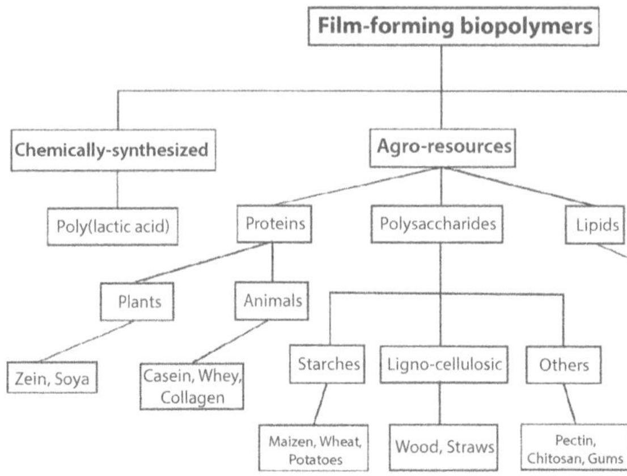

Figure 4.1 Classification of different film-forming biopolymers. Adapted from Garavand *et al.* [21].

non-biodegradable [19]. An alternative to these materials are polymers developed from natural sources, which are environmentally friendly.

Based on its source, the biodegradable biopolymers are grouped into three principal categories (Figure 4.1) [21, 24, 25] as follows:

1) chemically synthesized, applying monomers achieved from natural feedstocks, e.g., polylactic acid;
2) originating from agroindustrial resources—polysaccharides (starch), lignocellulosic materials (hocellulose, lignin, and derivatives), proteins (whey and collagen), lipids (bee wax), and free fatty acids;
3) obtainment by microbial fermentation: pullulan and polyhydroxyalkanoates.

The following section will highlight the structures, compositions, and properties of some raw materials that have the potential for biomedical applications.

4.2.1 Chitosan

Biopolymers are used in the biomedical, food, chemical, and industrial sectors [26–29]. One of the materials that have this ability is chitin polysaccharide; their principal derivative, chitosan, is more significant and plentiful marine polymer found worldwide. Its physical-chemical characteristics depend on the origin and obtaining method [26].

Chitin, mainly extracted from the exoskeleton of arthropods, is considered as the second most abundant polysaccharide found in nature. It is obtained in a solid form, insoluble in water, organic solvents, and dilute acid solutions. Chitin can be used as adsorbent and flocculant for water treatment. However, its applicability is mainly in chitosan, a polysaccharide obtained from the deacetylation of chitin (Figure 4.2) [30].

Chitosan, a de-N-acetylated analogue from chitin, is a heteropolysaccharide composed of linear β-1,4-linked units. It is noted that content and sequence of these polymer units can determine their physical-chemical and biological properties. During deacetylation for heterogeneous conditions, a blockwise distribution is observed; however, for homogeneous conditions, a random distribution of acetyl groups in chitosan is verified [31, 32].

The presence of hydroxyl and amino groups in chitosan makes it hydrophilic in nature. Chitosan can be used as a biomaterial because of its biocompatibility, biodegradability, non-toxicity, and mucoadhesive properties, essential for carrier systems of mucosal drugs [33].

Amino groups in chitosan are essential because besides producing hydrophilic chitosan, unlike chitin, they can be used in membranes and gels. These amino groups also allow for modification of chemicals in the structure of chitosan, besides giving an ionic character to the compound, increasing its applicability [34]. Thus, chitosan has affinity with various classes of compounds, such as herbicides, pesticides, dyes, lipids, and proteins. This enables its use in metal ion adsorption systems, bandages, and gels in carriers of drugs among others.

Another important property of chitosan, which distinguishes it from other polysaccharides, is its antimicrobial activity. Chitosan can inhibit the growth of microorganisms such as *Escherichia coli*, *Shigellady senteriae*, *Salmonella typhimuruim*, and *Candida*; this characteristic makes chitosan a highly important source of research for pharmaceutical and biomedical industries [35].

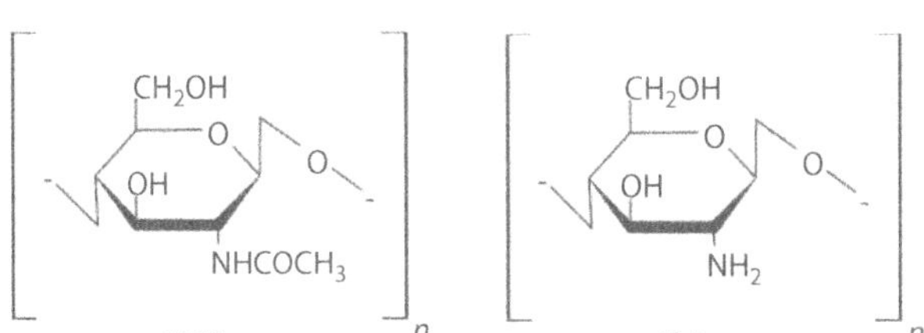

Figure 4.2 Representation of acetylated chitin [poly(N-acetyl-ß-D-glucosamine] and deacetylated chitosan [poly(D-glucosamine]. Adapted from Casadidio *et al.* [26].

Therefore, chitin and chitosan are produced on a large scale, and several established companies are exploring their potential in water treatment, cosmetics, food, health, agrochemicals, biotechnology, paper, and textiles, among others.

4.2.2 Alginate

Among various natural polymers, sodium alginate is a polysaccharide extracted from seaweed in the form of alginic acid, which reacts with a strong base (sodium hydroxide) to form sodium alginate salt.

The alginate extraction and purification processes are based on the conversion of an insoluble form, present in the cell wall of an alga, to a soluble form, usually sodium alginate [36]. For decades and till now, the extraction of algae has been studied for different applications. It is possible to obtain alginate particles with different physicochemical properties depending on gelation conditions and subsequent treatments. The use of an adsorbent of dry alginate beads facilitates storage and transportation, in addition to biodegradability, hydrophilicity, and stability in a wide pH range [37]. An alternative is its use in blend-forming processes with other materials, e.g., sericin, chitosan, pectin, poly(lactic acid), and others [38–41].

The residue from the extraction is usually alginate-based to produce animal feed, as it has not been explored as a bioadsorbent yet. Even after the extraction, it still contains much of the seaweed constituents. Economically, this fact makes the use of the residue as bioadsorbent interesting, because it reuses the wastes of the process.

Sodium alginate is a copolymer that has anionic chemical linear chain consisting of monosaccharide units of 1,4-ß-D-manuronic (M) and 1,4-α-L-guluronics (G), which are organized in the form of blocks [42, 43]. Residues can be organized into G (-G-)n, M (-M-)n, or alternate M and G (-MG-)n sequences or chains. Figure 4.3 shows these monosaccharide units in sodium alginate and how they organize themselves into alginate chemical chains.

The ratio of 1,4-ß-D-manuronic (M) and 1,4-α-L-guluronics (G) in the alginate directly affects the resulting films and gels, i.e., when the amount of 1,4-α-L-guluronic acid is more than that of 1,4-ß-D-manuronic acids, the films or gels get more rigid. However, if 1,4-ß-D-manuronic acid is present in a greater quantity in the alginate, the resulting films and gels will be more elastic [44].

The viscosity of the alginate is dependent on the sequential organization of blocks M and G. Table 4.1 shows the characteristics of alginates obtained

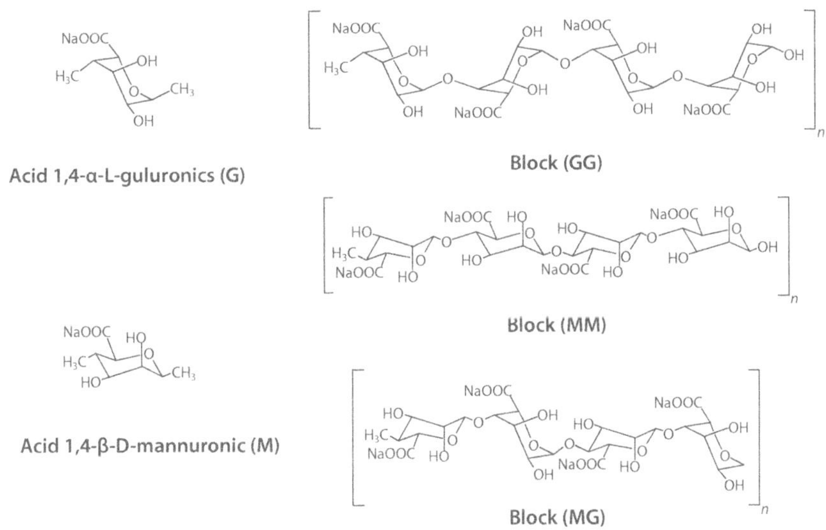

Figure 4.3 Structure of chemical chains alginate. Adapted from Hou and Wu [36].

from two main types of algae, one rich in group M and the other in group G. Commercially, they are divided into three classes based on viscosity: low viscosity (LV), medium viscosity (MV), and high viscosity (HV).

Alginate is commercially found in the form of salt, e.g., sodium alginate, potassium, and ammonium. Owing to its heavy colloidal properties: stabilizing and gelling agents; excellent biocompatibility, biodegradability, and non-toxicity; and low cost, alginate is considered a biopolymer and is largely employed in the food, pharmaceutical, textile printing, materials science, and other industries [36, 45–47].

In the attendance of divalent cations (Ca^{2+}), an aqueous solution of alginate can transform smoothly into 3D hydrogels, which gets it appropriate for drug delivery, scaffolds for tissue engineering, as well as models of extracellular matrices for biological investigations [36], and applications as a stabilizer agent for dental materials, regenerative medicine, and other [44, 48, 49].

Alginate also has a mucoadhesive property and can be used in curative membranes [50]. Momoh *et al.* [51] observed that alginate dressing is suitable as protein delivery system for wound healing. However, despite being widely used as a dressing for wounds, coating, and additives in the medical and food sectors, the alginate does not have any antimicrobial activity, acts as a selective barrier, which tracks humidity levels, and performs gas exchange with the environment among others [52].

Table 4.1 Chemical properties (XG and XM) and physical (μ, Mn, and dp) alginate.

Algae	Composition	Rheology	XG	XM	μ (dL/g)	Mn (kDa)	dp
Macrocystis pyrifera	High - M	LV	0.38	0.62	5.9	72.7	367
		MV	0.35	0.65	9.6	119.9	606
		HV	0.37	0.63	10.7	134.5	679
Laminaria hyperborea	High - G	LV	0.63	0.37	5.9	73.1	369
		MV	0.57	0.43	6.8	84.6	427
		HV	0.63	0.37	17.0	217.7	1,099

XG is the fraction of guluronic acid (G); *XM* is the fraction of mannuronic acid (M); *μ* (dL/g) is the intrinsic viscosity in 0.1 M NaCl; *Mn* (kDa) is the average molecular weight, and *dp* is the degree of polymerization.

4.2.3 Lignin

Lignin is a 3D macromolecule (aromatic) that is highly irregular in its constitution and amorphous, found in vegetables, associated with cellulose in the cell wall, adds rigidity and impermeability to water, and is resistant to microbiological and mechanical attacks (impact, compression, and fold) to plant tissues. This component, in general, because of its chemical structure, is more resistant to biological decomposition than other major biopolymers found in plant wastes (hemicellulose and cellulose) [53, 54].

Lignin is mainly distributed in combination with hemicellulose around cellulose strands in primary and secondary cell walls, and it is covalently bonded to the carbohydrate/cellulose structure (Figure 4.4a) [55, 56].

The chemical structure of lignin is on analysis for years but is not fully explored yet because it undergoes drastic changes during the wood insulation process. Moreover, lignin is a complex macromolecule formed by linked phenylpropanoid units. Phenylpropanoids comprise an aromatic ring attached to a propane group (three-carbon straight chain). These structures (Figure 4.4b) produce special alcohol molecules known as monolignols (*p*-cumarilic alcohols, *p*-coniferyl, and *p*-sinapilic), which are modified and bound together by polymerization reactions, dehydrogenated, and initiated by enzymes from three major structural units of lignin, namely, p-hydroxyphenyl, guaiacyl, and syringyl [57–59].

Depending on plant species and conditions under which it develops (regions, climatic conditions, and development stages), the structure of lignin can vary significantly. Basically, in hardwoods, such as eucalyptus, there is a predominance of guaiacyl and syringyl units, whereas in conifers, such as pinus, lignin is mainly composed of guaiacyl units, as grasses possess major proportions of p-hydroxyphenyl [58–60].

Several efforts have been made to evaluate the properties and applicability of lignin, mainly in the form of waste generated from the pulp and

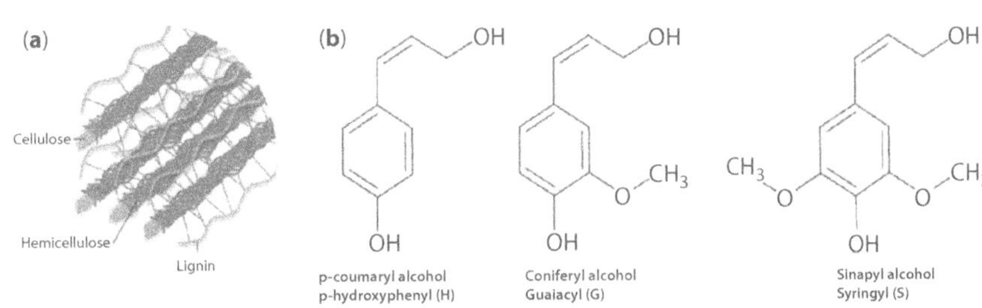

Figure 4.4 (a) Lignin distribution in lignocellulosic biomass and (b) lignin structure units. Adapted from Calvo-Flores and Dobado [57] and Figueiredo *et al.* [58].

paper industry. The polymer has a high stability, serving as an energy storage medium. Thus, lignin is usually burned in the aforementioned industry to reuse energy [61].

Lignin has a good ability to absorb heavy metal ions, given the phenolic groups, and can be potentially used in the treatment of sewage. Furthermore, it can be used to produce activated carbon because of its high carbon content. Activated charcoal is a porous material, which is widely used for adsorbing gases and effluent treatment [62].

The high carbon content makes lignin an important material to manufacture carbon fibers. Carbon fibers are mainly hard and low-density materials, which make them ideal to be used as reinforced materials in composites [62]. Usually, such fibers are made of polyacrylonitrile (PAN), which incurs a high cost of processing. The use of lignin can reduce cost, making it more viable, although the ways of obtaining lignin do not produce a pure material for economic viability.

Lignin has antioxidant and antibiotic properties [57], which can be used in feed for monogastric animals to aid in the development of intestinal flora. Another interesting product that can be obtained from lignin is vanillin, which is the primary component of vanilla extract.

Lignin can be used in many fields without changing its chemical structure, which not only makes the polymer more environmentally friendly but also reduces processing cost. It can be used in phenolic resins, polyolefins, polyesters, and polyurethanes [63]. To improve the incorporation of lignin polymer matrices, lignin's chemical structure can be modified, thus increasing the reactivity and solubility of it in organic solvents and facilitating its processing method [64]. This modified lignin can be used as cross-linking agent in epoxy resins and phenol-formaldehyde glues [57].

In most blends with thermoplastics, lignin becomes more brittle, affecting its viscoelastic properties [64]. Duval and Lawoko [62] studied the incorporation of Kraft-lignin, with other natural polymeric materials such as starch, soybean protein, and wheat gluten. This process allowed obtaining materials 100% bio-based, increasing hydrophylicity and rigidity.

4.2.4 Lactic Acid PLA

One of the less-complex acids is lactic acid (LA) (2-hydroxy propanoic acid), being composed by 2 (two) hydroxyl groups and 1 (one) chiral carbon atom [65]. LA has been majorly considered because of their use in the degradable bioplastics fabrication [66]. LA can be obtained by fermentation using suitable bacteria or a chemical route. It is the building block of PLA.

Poly (lactic acid) or polylactide (PLA) can be defined as a thermoplastic aliphatic polyester, which is extracted from renewable sources, for instance, cornstarch, sugarcane, and among other annually cultivated biomass products and wastes [67]. PLA is one of the most important polymers because of its large-scale production and use in the production of disposables such as cutlery, food jars, and packets. It is classified as a renewable synthetic as it is used as a precursor to lactide and is derived from starch [68].

For synthesizing PLA, two polymerization forms have been employed: (1) direct condensation polymerization of lactic acid monomers and (2) ring-opening polymerization of cyclic lactide dimmers [67]. Figure 4.5 shows the techniques, lactic acid monomers, and the feedstock of the fabricated PLA.

The PLA does not degrade readily in the soil, because of its rigidity at room temperature, and is limited to composting conditions. PLA degradation begins with hydrolysis on the polymer surface, reducing the molecular weight and release of CO_2, water, and other small molecules. The decomposition of the PLA compound can be enhanced by adding plasticizers [69–71].

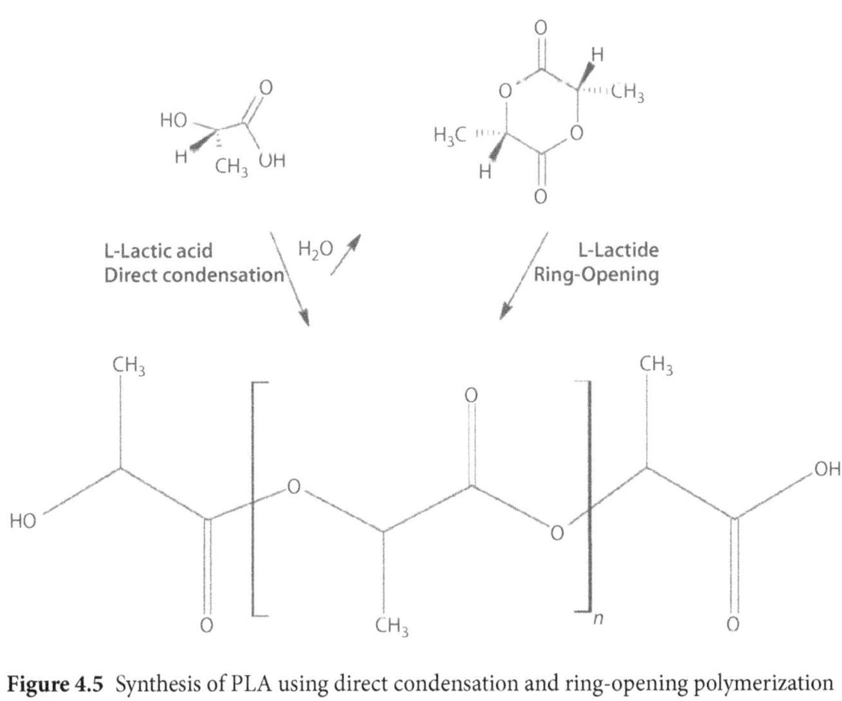

Figure 4.5 Synthesis of PLA using direct condensation and ring-opening polymerization of lactic acid and lactide, respectively. Adapted from Nofar *et al.* [67].

The authors investigated the decomposition shape of PLA bottles in a natural composting ambient at 65ºC for one month. It was verified that under actual composting conditions, the bottles were fully decomposed at 30 days. The decomposition procedure begun on the fourth day, and on the fifteenth day, a greater partitioning was performed [72]. Nevertheless, in other research, no decomposition of PLA films was noted in a garden composting after 6 weeks [73]. The more important is its brittleness characteristic, which minimize PLA from their widely mechanical requirements [65].

The PLA biopolymer provides a good environment for cell growth. As PLA is largely hydrophobic and has much less interaction with tissue cells when applied in tissue engineering, its applications are limited in the biomedical area [74]. Although the hydrophobic part permits the imprisonment of drug-loaded nanoparticles employing the mononuclear phagocyte process, the drug performance reduces when used *in vivo* [75]. Therefore, the bulk and surface modifications of PLA are required for their use in biomedical applications. An alternative is to mix with other polymers, for instance, PEO, PEG, polysaccharides, and polypeptides.

4.3 Properties Characterization of Green Adhesives for Biomedical Applications

In recent years, the biomedical industry and supporting researchers have widely studied some polysaccharides and potential for their possible application [48, 76]. In this study, a comprehensive overview of the general properties of green adhesives, such as alginate, chitosan, lignin, and polylactide, and their biomedical applications are presented, while novel opportunities for future researches with these natural polymers and/or polysaccharides are suggested.

Peltzer and co-workers [77] stated that in last years, the scientific community and civil society are concerned with the need for environmental conservation and are looking for new sustainable and renewable sources for developing biomaterials. Zhu *et al.* [78] proposed using advanced characterization methods, for example, *in situ* technologies and analysis of the physicochemical and structural changes of the active sites of lignocellulosic materials through spectroscopy.

In a study performed by Pulieri *et al.* [79], the effect of composition and cross-linking on the physical-chemical properties of biomedical samples were evaluated by scanning electron microscopy, thermogravimetry, contact angle measurements, dissolution, and swelling tests. The authors also investigated the mechanical properties of the chitosan samples, which were determined through stress-strain and creep-recovery tests. Carboxymethyl chitosan and chitosan-derived samples, whose biological and physicochemical properties are more enhanced than those of *in natura* chitosan, have emerged as promising candidates for different biomedical applications [80]. Additionally, within the biomedical context, carboxymethyl-based chitin and chitosan structures have been proposed as antimicrobial agents [80, 81], oxidants [80–83], drug delivery [84–89], and tissue regeneration platforms [80, 90, 91], owing to the exhibited biodegradability, biological activity, and low toxicity.

With the increasing use of biomaterials or green adhesives for biomedical applications, major developments are necessary to achieve tissue and disease specificity and even patient-oriented approaches [76]. Compared to other polymers, polysaccharides can undergo several complex chemical reactions, because these present an inherent variability of functional groups to be exploited for each desired aim [76]. Therefore, active compound systems of a film-food biopolymer, which presents a complex environment whose bioactivity is difficult to predict, were observed [92].

Different materials for biomedical applications, including green adhesives such as alginate hydrogels, have emerged owing to the ease of finely tuning their biochemical and viscoelastic characteristics, as well as their compliance with the cells' encapsulation [76, 93]. Nevertheless, alginate seems to be bioinert toward cells, because cells are apparently unable to adhere onto naturally occurring alginate isoforms [76].

Zhu *et al.* [78] used modern characterization technologies, namely, infrared spectroscopy, thermogravimetric analysis, scanning electron microscopy, transmission electron microscope (TEM), nuclear magnetic resonance (NMR), heteronuclear single quantum coherence techniques, and gas chromatography mass spectrometry, to clearly and efficiently analyze the precise structure of lignin and its derived compounds, which are extremely important to biomedical applications.

A series of cross-linked biomaterials with different compositions was produced and their physicochemical properties were investigated by diffraction X-Ray, atomic force microscopy, scanning electron microscopy, wettability, Fourier transform infrared, inductively coupled-optical emission spectrometry, and surface area-porosimetry analyzer; besides mechanical (stress-strain) and thermal behavior (thermogravimetry analysis), which were compared to the corresponding uncross-linked samples [77, 78]. For this reason, authors as Peltzer *et al.* [77] and Zhu *et al.* [78] also performed a complete characterization of the biodegradable films

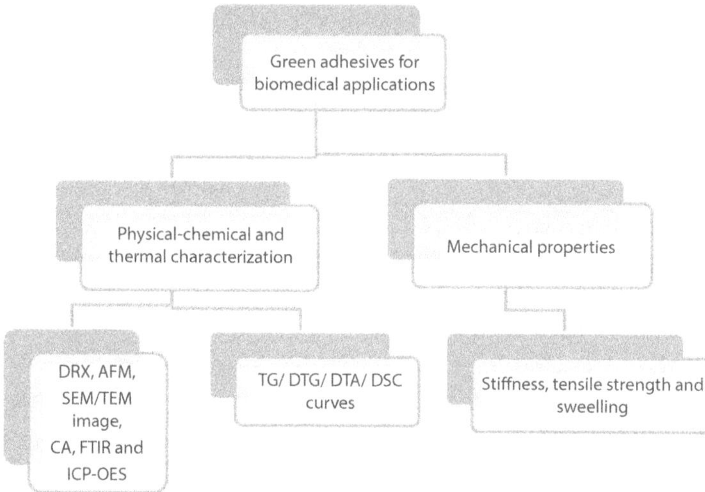

Figure 4.6 Schematic representation of the main physical-chemical characterization, thermal behavior, and mechanical properties applied to green adhesives for biomedical applications.

based on physicochemical characterization, structural properties, tensile properties, water vapor permeability, resistance to water, and thermal stability. The following sections presented the main characterization techniques commonly used to measure the physical-chemical and mechanical properties, and thermal behavior of the green adhesives, when these biomaterials are applied in industry and/or in biomedical research (Figure 4.6).

4.3.1 Diffraction X-Rays (DRX)

Diffraction X-Rays evaluated the main physical properties, i.e., amorphous and crystalline regions of some green adhesives used in biomedical applications. For example, polylactides properties were related to the enantiomeric purity of the lactic acid stereo copolymers [94]. Copolymers, such as poly-(D-lactide) or poly-(L-lactide) and high D-lactide or L-lactidecopolymers, presented significantly ordered structures and crystallized. However, large molecular weight polylactides were either amorphous or semicrystalline at ambient temperature, depending on the amounts of L-, D-, and meso-lactides in the structure [94]. According to Tsuji and Ikada [95], PLA structure can be fully unstructured or exhibit until 40% crystallinity. Conversely, PLA resins, which contain above 93% of L-lactic acid, are semi crystalline while PLA, containing 50%–93% L-lactic acid, is narrowly amorphous.

The occurrence of both meso- and D-lactide forms produces imperfections in the crystalline structure, decreasing the percentage crystallinity [94]. Another green adhesive analyzed by X-Ray diffraction patterns was chitosan, which exhibited a polymorphic nature, occurring in three different crystalline forms named asα-, β-, and γ-chitosan [96, 97]. Furthermore, by X-Ray diffraction, the *in natura* lignin exhibited similar patterns to the derived-lignin compounds. The results demonstrated high efficiency and stability of lignin and affirmed presence of amorphous character [78]. Rinaudo [98] affirmed that in solid state, X-Ray analysis of the extended biofibres suggested a helical conformation for the different types of blocks and enabled the identification of the interaction with counter ions as well as the change in this conformation. For the alginate green adhesives, the X-Ray pattern displayed broad and sharp peaks, signifying the presence of microcrystalline structures, due to intra-hydrogen bonding [99, 100].

4.3.2 Atomic Force Microscopy (AFM)

According to Markert *et al.* [101] and Chang *et al.* [102], the atomic force microscope (AFM) is a powerful tool that characterizes localized mechanical properties (mapping topographical and roughness) of both cell culture substrates and biological samples. This analysis requires the selection of a contact point (CP) and is often time consuming and subjective due to the influence from intermolecular forces and low signal-to-noise ratios (SNR) [101, 102].

For the green adhesives, inorganic nanoparticles, and/or biomedical samples, the dimensions of the stabilizer layers can be estimated by the difference in size of the nanoparticles obtained by the AFM images [103]. AFM analysis demonstrated that the biofilm surfaces appeared granular, possibly due to the presence of the remaining cell walls [77].

In their study, Chen *et al.* [104] investigated the surface stiffness of chitosan-based biofilms by AFM. More information on the surface morphology of the bioadhesives were achieved by means of AFM topographic pictures. The authors observed that fresh chitosan bioadhesives presented a fairly low level of surface roughness, while the modified chitosan biofilms showed rougher surfaces and contained some orifices (dark regions). These hollows were possibly the result of the imperfections created in the bioadhesives caused by cross-linking of the modified chitosan molecules.

For Raphey *et al.* [105], modifications in the tweezers whereas operating the AFM involved certain phases leading to different procedures. An AFM figure was chosen for the tweezers closed and a particular voltage was applied between the biofilm, when the adhesion between the biofilm and nanosubstance was larger than between the nanosubstance and substrate, the material could be expelled by applying a potential difference between the nanosubstance and the substrate. Additionally, chitosan-based nanoparticles (NPs) of a mucoadhesive polymer for biomedical applications were produced and subjected to the AFM technique for evaluation of their morphology and texture. The authors confirmed the size of the prepared NPs, which exhibited a size range from 80 to 100 nm. From the obtained images, it became clear that the prepared NPs presented spherical morphology [106].

Another study conducted by Rao *et al.* [107], also using AFM, revealed the morphological properties of biomedical samples, which presented the surface morphology with smooth spherical shapes, nano-range sizes, particle size distribution, and stabilized multi-functional groups. It is important to highlight that the greater size distribution can be attributed to the variations in the number of molecules being attached to nanoparticles [99, 107].

4.3.3 Scanning Electron Microscope (SEM Images)

Siqueira *et al.* [108] used SEM images to morphologically characterize (surface structure) the alginate samples for biomedical applications. Such a technique enabled the observation of the properties that may directly affect the cell growth within such biomaterials, as the properties play an important role on the transport of nutrients and other biological fluids. The authors demonstrated the effect on the morphological properties of the alginate-based gels caused by the addition of nanocelluloses into them. The micrographs presented a typical morphology of freeze-dried biomaterials with a high porosity [109]. It was also observed through SEM images, that addition of nanocelluloses apparently contributed to the formation of pores with larger diameter than those of pure cross-linked alginate gels [108]. However, for cross-linked and uncross-linked samples, phase separation could not be detected via the SEM analysis of fractured sections, suggesting good biocompatibility of the chitosan or other green adhesives in the compositions [79].

Another study, Lee and Mooney [48] used different microscopy types and magnifications, such as optical, confocal, scanning electron, and transmission, and revealed the more appealing characteristics of alginate biofilms for some biomedical utilizations, including biocompatibility, mild gelation conditions, adhesion and proliferation of myoblasts cultured on alginate gels, formation of growth-plate-like structures, which could potentially substitute dysfunctional epiphyses, and simple modifications to prepare alginate derivatives with new biological properties.

Park *et al.* [110] used SEM micrographics to investigate the morphology of the PLA biofilms and their composites for biomedical applications. These images revealed that the scale and form of the spherule depended on the starch amount of the composite. As the starch quantity augmented, the spherulite size reduced and became most irregular, suggesting that starch can play a role as a nucleating factor in the PLA biofilms. The SEM images showed good adhesion of the starch grains to the PLA matrix up to 5% starch amount. As the starch quantity raised, holes caused by the separation of the starch in the matrix proceeded [110].

The SEM images of PLA biofilms for green adhesives exhibited a relatively smooth surface, without significant defects and a homogeneous morphology aspect, exhibiting uniform distribution and relatively good interfacial adhesion, caused by relatively strong interaction between components [111]. However, incorporating chitosan powder especially at high concentrations led to significant changes in morphology whereby due to the phase separation, the bio composites exhibited a non-homogeneous morphology with some agglomeration areas, which was evident at high magnifications, as well as the presence of the chitosan in the polymeric matrix, causing an uneven surface [111].

Larreñeta *et al.* [112] also used SEM images to characterize lignin green adhesives and verified that lignin-based hydrogels demonstrated more important endurance to bacterial adherence in comparison with polyvinyl chloride biofilms and resembling hydrogels that did not contain lignin. Accordingly, lignin-based hydrogels presented encouraging features as biomedical material coverings based on their resistance to contagion and capacity to release drugs over several days.

4.3.4 Wettability or Contact Angle (CA)

Surfaces hydrophobic or hydrophilic modifications are commonly required, when green adhesives are applied in several biomedical treatments, which can be evaluated by wettability or contact angle. For instance, introducing hydrophobic units to chitosan enhanced its binding to cell surfaces, facilitated the DNA dissociation inside the cell, protected the treatment from enzymatic degradation, and alleviated serum inhibition leading to favorable gene transfection [113]. Kalantari *et al.* [114] revised novel progresses in the biomedicine industry of ecologically correct (green adhesives) chitosan and chitosan-mixture electrospun nanofibers. Results of this technique publicized that manufactured scaffolds can potentially be used in clinical exercise for wound treatment. Based on their discoveries, the incorporation of insulin containing chitosan particles improved the hydrophilicity, water-uptake, and blood biocompatibility.

In hydrophobic matrices, the better compound compatibility leads to polymer plasticization, reducing both stiffness and strength, but enhancing the biofilm extensibility [92]. For Wei *et al.* [93], the different lignocellulosic materials, especially lignin, are often investigated by wettability or contact angle to predict the existence of hydrophilic (absorbed water) and hydrophobic (repulsed water) groups. These are the main characteristics responsible for biocompatibility, biodegradability, low thermal conductivity, thermal insulation, and some mechanical properties, such as adhesion phenomenon, viscosity, and surface recovery [14, 115, 116].

For Pulieri *et al.* [79] and Saltzman [117], several parameters can contribute for the cell adhesion and this can be enhanced on surfaces with the help of intermediary wettability. Analyzed samples exhibited a lightly hydrophilic performance with contact angles ranging between 78° and 90°. Dehydrothermal (DHT) cross-linking only slightly reduced the wettability of gelatin and the blend samples, possibly because of the formation of amide and ester bonds, decreasing the content of the most hydrophilic free carboxyl, amine, and hydroxyl portions [79]. Conversely, chitosan hydrophilicity slightly increased because of DHT treatment, indicating a macromolecular reorganization, leading to the preferential orientation of the more hydrophilic portions on the surroundings [79]. Nevertheless, the authors observed that contact angles of cross-linked and uncross-linked samples did not change considerably for studied samples, while differences in surface wettability were evidenced for the other cross-linked and uncross-linked mixture samples [79].

4.3.5 Fourier Transform Infrared Spectroscopy (FTIR)

Another important technique used in the characterization of green adhesives for biomedical applications is the Fourier transform infrared (FTIR), which shows the principal absorption bands in the infrared region. Infrared (IR) radiation broadly refers to the portion of the electromagnetic spectrum between the visible and the microwave region (i.e., 2,500 and 15,000 nm, respectively) [94]. For PLA biopolymers, it was possible to observe maximum absorbance, which can be attributed to the ester group present in the helical backbone of several marine species [94, 118]. Kister *et al.* [118] identified some IR regions for PLA biofilms. For example, the strong bands at 2,997, 2,946, and 2,877 cm^{-1} were assigned to the CH stretching region, and the CH_3 and CH modes. A large band of the C-O stretching region was observed at 1,748 cm^{-1}. The region between 1,500 and 1,360 cm^{-1} was characterized by the CH_3 band. The CH deformation and asymmetric bands appeared at 1,382 cm^{-1} and 1,365 cm^{-1}, respectively. The bands that appeared at 1,315 and 1,300 cm^{-1} were due to the CH bending modes. In the region from 1,300 to 1,000 cm^{-1}, one could observe the C-O stretching modes of the ester groups at 1,225 cm^{-1} and the O-C asymmetric mode at 1,090 cm^{-1}. Between 1,000 and 800 cm^{-1}, peaks could be observed at 956 and 921 cm^{-1}, attributed to the characteristic vibrations of the helical backbone with CH_3 rocking modes. Two bands relating to the crystalline and amorphous phases of PLA were observed at 871 and 756 cm^{-1}. The peak at 871 cm^{-1} could be assigned to the amorphous phase and the peak at 756 cm^{-1} to the crystalline phase. The peaks below 300 cm^{-1} mainly corresponded to the CH_3 torsion modes and the skeletal C-C torsions [94].

4.3.6 Inductively Coupled Plasma—Optical Emission Spectrometry (ICP-OES)

Over current years, reinforced polylactic acid (PLA) composites and other green adhesives have been investigated as possible biocompatible and bioabsorbable materials for biomedical applications [111, 119]. In a recent study conducted by Ferrández-Montero *et al.* [119], inductively coupled plasma-optical emission spectrometry (ICP-OES) was employed as a new characterization technique to measure the ion concentration (inorganic and/or metallic compounds) released into the biological ambient. The authors demonstrated that microparticles added to a biodegradable polymer could solve the absence of bioactivity and the few mechanical properties presented by the polymers [111]. Additionally, they could neutralize the harmful effects related with the high degradation rate of chemical elements, such as alkalinization and prominent H_2 release [119]. They also observed that the successful dispersion of microparticles was achieved by introducing a colloidal phase in the processing, where organic additives were introduced to change the particle surface and further a chemically stable suspension [119]. The resulting colloidal suspension was then used as raw material to achieve composite biofilms by melting process. The biofilms exhibited advantageous *in vitro* behavior in terms of degradation, hydrogen release, and oxygen permeability [111]. Moreover, the viability with fibroblast cells opens a window of opportunity for these composite biofilms as bioabsorbable materials for tissue engineering and wound dressing applications [119]. It is also worth commending this specific technique that characterizes green adhesives for biomedical applications, because few studies reporting its applications exist in the literature [111, 119].

4.3.7 Thermal Analysis (TG/DTG/DTA and DSC Curves)

In Pulieri *et al.* [79], TG/DTG, DTA, and DSC curves were used to verify the main steps of thermal degradation that samples of green adhesives had undergone when employed in biomedical applications. For example, Siqueira *et al.* [108] performed thermogravimetric analyses of the alginate gels to obtain information about the thermal stability of the materials based on their compositions (with and without nanocelluloses). Based on the thermogravimetric curves, it was possible to compare different behaviors of the gels and analyze as the effects of treatments as the cross-linking density or oxidation of nanocelluloses influenced the final properties of the gels, which determined the following as the three major mass loss events: (1) loss of moisture and other volatile components; (2) complex degradation of glycoside chains, for example, depolymerization, dehydration, and thermal decomposition of glycosidic units; and (3) degradation of carbon-based residues, and formation of volatile products with low molar weight [108]. According to Han *et al.* [120], the residue from the thermal decomposition can be assigned to the electrostatic interaction between the alginate and nanocellulose with calcium ions. This results in strong thermal stability that is predominantly detected in biocomposite samples [120]. The influence of nanocellulose amounts on the thermal stability of alginate gels was analyzed, and it was observed that a lower nanocellulose concentration in the alginate gels has few influences on its final properties. This event is fewer evident for gels containing cellulose nanofibers, depending on the lower dispersion of the modifications of these in the alginate matrix [108].

Ferrández-Montero *et al.* [119] studied the thermal behavior of as-received PLA pellets and biofilms by TGA analysis conducted from room temperature to 500°C in an argon atmosphere. The TG/DTG measurements primarily provided information about PLA modification due to the addition of magnesium and real inorganic portion of the composites. In addition, this method proved that there are no identifiable solvent traces or impurities in the samples of composites. The TG/DTA curves illustrate the initial degradation, maximum, and burnout temperatures, also mass loss at the first peak and the percentage of magnesium for the pure biopolymer and its composites.

Furthermore, PLA biofilms have lower melting point and glass transition temperatures when compared with polyethylene terephthalate (PET) and polystyrene (PS) films. The glass transition temperature of PLA undergoes variations over time [94]. Moreover, moisture content between 10% and 95% and storage temperatures from 5°C to 40°C do not present an effect on the transition temperature of PLA biofilms [94]. This may be described by its little water sorption values. PLA seals well at temperatures below the melting point; however, a considerable shrinking of the films has been verified when the biomaterial is close its melting point [94].

4.3.8 Surface Area and Porosimetry Analyzer (ASAP)

Koo *et al.* [121] affirmed that lignocellulosic materials showed pores, which can be generated due to the biological, chemical, and mechanical treatments. They evaluated that generation of pores increased both the specific surface area and enzymatic access into the interstices of the different bio-substrates [115]. The surface area and porosimetry analysis was also applied by several researchers to evaluate green adhesives for biomedical applications. Raphey *et al.* [105] related that with advances in nanotechnology, the applications of nanomaterial are increasing rapidly. The large surface area of green biofilms and the ability to handle their surfaces and dimensions has increased their potential application. Bioadhesives possess a larger potential for applications in biomedicine due to their electrical characteristics, biological, chemical, thermal behavior, and mechanical properties [105]. The main reason referent to the toxicity of carbon nanotubes is their minimum size, which causes an increase in their total surface area. Moreover, this suggests either that the contact area with the cellular membranes has been raised or that the absorption/transportation of toxins has been substantially improved [122, 123]. The increased surface area of bioadhesives makes them a major vehicle for the detection of biological species. The enzymatic and electrochemical response of biological sensors has enhanced after the incorporation of the surfactant-modified biofilms because of its high surface area networks [105].

Zhu *et al.* [78] revised various works that deal with physical-chemical properties from the extraction of *in natura* lignin and the several thermal pretreatment processes of this lignocellulosic material for biomedical and industrial applications. In this investigation, the authors identified the developments of porosity (e.g., low, moderate, and high), specific surface area (low, intermediate, and high), pore formation (start, nucleation, and active sites), pore volume (size and distribution), and pore diameter (micro, meso, and macropores).

4.3.9 Mechanical Properties of Green Adhesives

The design and production of new biopolymers or green adhesives for biomedical applications with superior mechanical properties can potentially be achieved using genetic engineering techniques that control bacterial synthesis [48]. For Chen *et al.* [104], biofilms manufactured from raw chitosan tend to be more strong, hard, and brittle. Therefore, plasticizers are needed to reduce the frictional forces between the polymer chains and thereby better their mechanical properties. Other hydrogels such as alginate gels have low mechanical stiffness and more general physical properties [48]. However, matching the physical properties of alginate gels to the particular needs of an application is a continuing challenge. Gels suitable for each application can be produced by considering a range of different cross-linking strategies, molecules with several chemistry structures, molecular weights, and cross-linking functionality [48].

It was found that the tensile properties of the PLA-based biocomposites vary depending on the amount of chitosan added. It is clear that an increase in the quantity of chitosan in the plasticized PLA increases Young's modulus and decreases both tensile strength and elongation at break [124]. These improvements can be attributed to well-dispersed chitosan and attractive interactions between the components of the biocomposites. Bijarimi *et al.* [125] highlighted that tensile and flexural strength, stiffness and notched Izod impact strength decreased significantly when the PEG (polyethylene glycol) concentrations were added to the PLA matrix. It was found that the PLA/chitosan composite material showed the appropriate porosity and structure that enabled it to retain certain shapes and mechanical properties [126]. The processing morphological, structural, thermal, and mechanical performance of electrospun biocomposites based on PLA blended with chitosan or other microparticles have been studied by Arrieta *et al.* [126]. Chitosan creates point defects in the fibers, which causes a reduction in the mechanical behavior of biocomposites. On the other hand, antioxidants improve the thermal stability of biocomposites and produce irregularities-free fibers with better mechanical performance [126].

Finally, nature-based biocomposites can be used to engineer new classes of biofilms or green adhesives with precisely controlled biological, chemical, physical, and mechanical characteristics, unlike the limited selection available from natural sources. This could revolutionize the use of these materials for different biomedical applications.

4.4 Biomedical Applications of Natural Polymers

The medical and biomedical use of polymers is a very important and continually growing area of research. Of the available natural resources, biopolymers sourced from the ocean have been widely used because they are

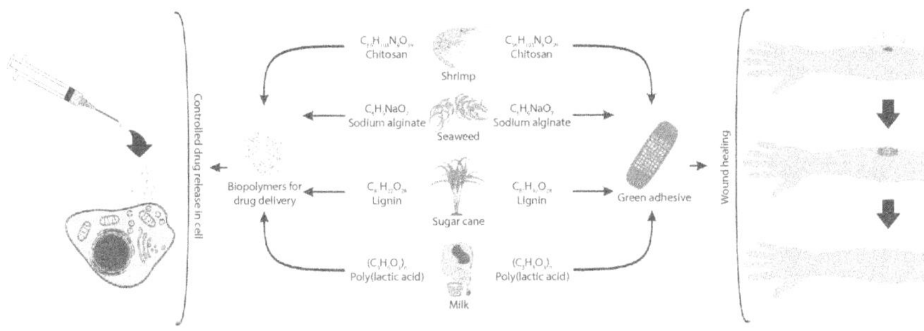

Figure 4.7 Schematic representation of the biomedical applications using alginate, chitosan, lignin, and PLA (polylactic acid).

free of communicable diseases in addition to allowing high production at low cost [127, 128].

Among the natural polymers, alginate, chitosan, lignin, and PLA (polylactic acid) have several biomedical applications (Figure 4.7).

4.4.1 Alginate

Alginate is a polysaccharide derived from brown algal metabolism [108]. It is a linear polysaccharide copolymer with anionic carboxylic groups which are easily substituted by monovalent sodium cations [129]. Its bioactivity and physicochemical adaptability have been studied and applied in various biomedical fields such as regenerative medicine, tissue engineering, and pharmaceuticals [128].

4.4.1.1 Biomedical Applications of Alginate

Alginate is known for its mucoadhesive properties that can be used to prolong drug delivery time on mucosal surfaces and thereby improve drug absorption [130]. Some alginate hydrogels are electrostatically interactive *in situ* and can be used in drug delivery systems to treat certain diseases [129].

Hydrogels have the properties of 3D polymer networks and a natural extracellular matrix (NECM) and have been widely studied. Calcium alginate polysaccharide hydrogel has a wide range of functionalities as well as features suitable for cell envelopment and tissue regeneration [108, 128, 131].

Both synthetic and naturally occurring hydrogels have facilitated the control of some physical properties such as mechanical strength, swelling, degradation, and stiffness in order to stimulate cells to form functional tissues. Bioactive polymers extracted from natural sources have shown superior biofunctionality compared to synthetic polymers. In addition, they have prompted low immune response and optimum biocompatibility in biomedical applications [128].

Three-dimensional bioprinting is widely used in tissue engineering and regenerative medicine to develop complex organic and artificial structures that mimic native organs and tissues [132]. Moreover, the US Food and Drug Administration (FDA) approved alginate for human use in healing wounds [131]. Therefore, alginates are used as dressing materials, mainly sodium alginate in the form of a hydrogel [127].

Alginates can be used in bead, capsule, and hydrogel forms for the immobilization of enzymes primarily through encapsulation. They are biocompatible, water soluble, less toxic, and relatively inexpensive compared to commercially available products [129].

4.4.2 Chitosan

Chitosan is a biopolymer used in biomedical applications. It is a favorable material for current pharmaceutical formulations due to its ability to improve the solubility and stability of bioactive molecules. It also has mucoadhesive, antioxidant, analgesic, hemostatic, non-immunogenic, and antimicrobial effects [111, 133].

4.4.2.1 Biomedical Applications of Chitosan

Chitosan-based nanoparticles can inhibit tumor cell growth by inducing apoptosis and presenting high permeability. Particles up to 100–200 nm in size can be absorbed by receptor-mediated endocytosis, while larger particles should be upbrought in by phagocytosis. Chitosan, in its pure form or mixed with other polymers, active agents, and metallic nanocomposites, has been widely used in many biomedical applications. This includes drug delivery, where it is used as a nanometric vehicle for inhibiting tumor tissue growth while minimally affecting normal tissue sites [134].

Chitosan has antioxidant properties that correspond to its ability to eliminate different species of oxygen radicals such as superoxide, hydroxyl, DPPH (2,2-diphenyl-1-picrylhydrazyl) alkyl [26].

Chitosan is an excellent choice as a healing material because it has intrinsic antimicrobial properties and provides extrinsic antimicrobial agents to the infected area. Hydrogels are characterized by their ability to store high volumes of water within their structure. Therefore, they can be used to moisten the infected area [134].

Chitosan has the ability to inhibit reactive oxygen species (ROS) by donating a hydrogen or electron pair. The following mechanisms explain the antibacterial and antifungal properties of chitosan: 1) positively charged chitosan interacts with the negatively charged groups on the surface of microbial cell membranes, which in turn alters their permeability and results in osmotic imbalance; 2) chitosan binds with cellular DNA, and inhibits microbial RNA synthesis; and 3) caused by spore element suppression, and metal chelation and binding to nutrients essential for microbial growth [111].

Chitosan is known as an active agent that improves the efficiency of antibiotics. Furthermore, recent pharmaceutical formulations report the use of this biopolymer in fighting resistant microorganisms [133].

Also, chitosan shows expressive osteoconductivity and minimal osteoinductivity. Its causes *in vivo* osteoblast proliferation and neovascularization [23].

4.4.3 Lignin

Lignin has been reported as a biomacromolecule with antioxidant and antimicrobial properties. Therefore, it has potential to be used in biomedical applications [112].

4.4.3.1 Biomedical Applications of Lignin

Hydrogels are mainly composed of homopolymers or copolymers because of chemical or physical cross-linking. Furthermore, they are commonly used in a variety of applications including scaffolding in tissue engineering, drug release carriers, cartilage repair, 3D cell culture, and smart devices [135].

Because of the antimicrobial characteristics of lignin, the hydrogels of this polymer have been evaluated and shown to be potential solutions for fighting biofilms on medical devices because they can be manufactured as antimicrobial coatings [112].

Both lignin and chitin have important properties such as antioxidant and antimicrobial activity, providing mechanical strength, stiffness, and protection against ultraviolet radiation. In addition to their important properties in plants and animals (crustaceans, mollusks, and insects), their sustainable nature has inspired researchers, as both lignin derivatives are used in tissue engineering, antioxidant absorption, antimicrobial activity, diabetes treatments, dressings, and other biomedical applications [96].

4.4.4 Polylactide (PLA)

Polylactide (PLA) and polylactic-co-glycolic acid (PLGA) are biodegradable synthetic polymers that are of potential biomedical interest due to properties such as biocompatibility, biodegradability, nontoxicity, safe elimination, and tunable mechanical and physicochemical characteristics. In addition, there is the possibility of obtaining novel polymeric materials from their chemical properties [136, 137]. The main biomedical application of PLA and PLGA is as scaffold materials in tissue engineering [137], for the development of drug delivery systems, and for the synthesis of polymeric nanofibers (NFs) using the electrospun technique, which can potentially improve therapy or diagnostic of cancer, for example [138].

4.4.4.1 Biomedical Applications of PLA

Doxorubicin (Dox) is a hydrophilic chemotherapeutic agent used to treat liver, breast, ovarian, and lung cancer, but, its usage decreases the white blood cell count and causes alopecia and cardiopaties. In this context, doxorubicin encapsulated through polylactic acid (PLA) has specific potential applications such as moving to a desired location and triggering drug delivery [139].

PLA nanofibers and gold (Au) nanoparticles as well as PLA/Au nanocomposites can potentially be used as drug carriers for efficient drug delivery to the target cancer cell. Moreover, the PLA/Au nanocomposites conjugated with daunorubicin have shown a synergistic effect on both sensitive and drug-resistant leukemia cells by facilitating either the cellular absorption of the chemotherapeutic agent into drug-resistant leukemia cells or inhibition of the multiplication of cancer cells [140].

The targeted nanomedicine uses macrophages as theranostic nanoparticles (NPs) from the polymer biodegradable photoluminescent poly (lactic acid) (BPLP-PLA). Moreover, it can be used to reduce the toxicity of chemotherapy to human cells. These nanoparticles that were internalized within macrophages showed strong fluorescence, biodegradability, and cytocompatibility. The immune cell mediated drug delivery system enables self-powered, effective, targeted, and traceable cancer drug delivery [141].

Electrospun nanofibers prepared with Poly (D,L-lactide-co-glycolide) (PLGA) are important in mediating many biological actions, such as adhesion to osteoblasts. They assist the survival and proliferation of human mesenchymal stem cells in osteoblasts and chondrocytes. In addition, this formulation promotes adhesion of interstitial and endothelial substances [142].

Poly (lactic acid) (PLA)-based biomaterials may be used for many orthopedic regenerative engineering applications by regenerating various orthopedic tissues, including bones, ligaments, cartilage, and menisci [143]. Graphene-PLA nanocomposites with higher-order 3D geometry can be used for surgical reconstruction and drug delivery [144].

4.5 Final Considerations

The use of biomaterials obtained from different renewable sources is both innovative and promising, with a growing demand in both biomedical research and in industry. In this chapter, the use of green adhesives for biomedical applications was presented and discussed as a viable alternative to new synthetic bioadhesives. Further, four possible biomaterials, namely, alginate, chitosan, lignin, and PLA, and their blends that were extracted from renewable sources were presented. Moreover, their composition, structure, and the different characterization techniques that are currently employed to evaluate and obtain their chemical, physical, biological, and thermal behavior and mechanical properties were discussed. Furthermore, by referencing several studies, it was shown that there are wide applications in medicine for bioadhesives obtained from these renewable sources.

Some research on the performance of green adhesives for different applications within the biomedical field has shown that the use of natural polysaccharides and polymers as a feedstock for bioadhesives is feasible because the performance observed for the different applications is satisfactory. Therefore, it can be said that the use of green adhesives to replace non-renewable polymers with natural-resource polymers as feedstock for bioadhesives is a vital milestone for the sustainable progress of these modern biomaterials.

Acknowledgements

The authors gratefully acknowledge Foundation for Research Support, Scientific and Technological Development of Maranhão - FAPEMA (01190/17) and National Council for Scientific and Technological Development - CNPq (426162/2018-8) for the financial support. They also thank the Ceuma University and Federal University of Maranhão for technical and professional support.

References

1. Donkerwolcke, M., Burny, F., Muster, D., Tissues and bone adhesives - historical aspects. *Biomaterials*, 19, 1461, 1998.
2. Maitz, M.F., Applications of synthetic polymers in clinical medicine. *Biosurf. Biotribol.*, 1, 161, 2015.
3. Assmann, A. *et al.*, A highly adhesive and naturally derived sealant. *Biomaterials*, 140, 115, 2017.
4. Spotnitz, W.D. and Burks, S., Hemostats, sealants, and adhesives: Components of the surgical toolbox. *Transfusion*, 48, 1502, 2008.
5. Balakrishnan, B. *et al.*, A novel injectable tissue adhesive based on oxidized dextran and chitosan. *Acta Biomater.*, 53, 343, 2017.
6. Green, J.J. and Elisseeff, J.H., Mimicking biological functionality with polymers for biomedical applications. *Nature*, 540, 386, 2016.
7. Ishihara, M. *et al.*, Polyelectrolyte complexes of natural polymers and their biomedical applications. *Polymers*, 11, 672, 2019.
8. Preethi, G.U. *et al.*, Biomedical applications of natural polymer based nanofibrous scaffolds. *Int. J. Med. Nano Res.*, 2, 2, 2015.
9. Zhu, W., Chuah, Y.J., Wang, D., Bioadhesives for internal medical applications: A review. *Acta Biomater.*, 74, 1, 2018.
10. Marshall, S.J. *et al.*, A review of adhesion science. *Dent. Mater.*, 26, e11, 2010.
11. Villanueva, J.G.V. *et al.*, Bio-adhesion evaluation of a chitosan-based bone bio-adhesive. *Int. J. Adhes. Adhes.*, 92, 80, 2019.
12. Guo, J. *et al.*, Development of tannin-inspired antimicrobial bioadhesives. *Acta Biomater.*, 72, 35, 2018.
13. Laffleur, F. and Küppers, P., Adhesive alginate for buccal delivery in aphthous stomatitis. *Carbohydr. Res.*, 477, 51, 2019.
14. Zain, N.M. *et al.*, Effect of surface treatments on the durability of green polyurethane adhesive bonded aluminium alloy. *Int. J. Adhes. Adhes.*, 55, 43–55, 2014.
15. Oliveira, S.M., Alves, M.N., Mano, J.F., Cell Interactions with superhydrophilic and superhydrophobic surface. *J. Adhes. Sci. Technol.*, 28, 843, 2014.
16. Scognamiglio, F. *et al.*, Adhesive and sealant interfaces for general surgery applications. *J. Biomed. Mater. Res. Part B*, 104, 626, 2016.
17. Khanlari, S. and Dubé, M.A., Bioadhesives: A Review. *Macromol. React. Eng.*, 7, 573, 2013.
18. George, B. and Suchithra, T.V., Plant-derived bioadhesives for wound dressing and drug delivery system. *Fitoterapia*, 137, 104, 2019.

19. Khanzadi, M. *et al.*, Physical and mechanical properties in biodegradable films of wheyprotein concentrate-pullulan by application of beeswax. *Carbohydr. Polym.*, 118, 24, 2015.
20. Morin-Crini, N. *et al.*, Applications of chitosan in food, pharmaceuticals, medicine, cosmetics, agriculture, textiles, pulp and paper, biotechnology, and environmental chemistry. *Environ. Chem. Lett.*, 17, 1667–1692, 2019.
21. Garavand, F. *et al.*, Improving the integrity of natural biopolymer films used in food packaging by crosslinking approach: A review. *Int. J. Biol. Macromol.*, 104, 687, 2017.
22. Bilal, M. and Iqbal, H.M.N., Naturally-derived biopolymers: Potential platforms for enzyme immobilization. *Int. J. Biol. Macromol.*, 130, 462, 2019.
23. Bernal-Ballen, A., Lopez-Garcia, J.A., Ozaltin, K., (PVA/Chitosan/Fucoidan)-Ampicillin: A Bioartificial Polymeric Material with Combined Properties in Cell Regeneration and Potential Antibacterial Features. *Polymers*, 11, 1325, 2019.
24. Tang, X., Kumar, P., Alavi, S., Sandeep, K., Recent advances in biopolymers and biopolymer-based nanocomposites for food packaging materials. *Crit. Rev. Food Sci. Nutr.*, 52, 426, 2012.
25. Jafari, S.M. *et al.*, Hydrophobicity, thermal and micro-structural properties of whey protein concentrate-pullulan-bees wax films. *Int. J. Biol. Macromol.*, 80, 506, 2015.
26. Casadidio, C. *et al.*, Chitin and chitosans: Characteristics, eco-friendly processes, and applications in cosmetic science. *Mar. Drugs*, 17, 369, 2019.
27. de Souza, K.V. *et al.*, Use of biopolymer in pulp and paper industry wastewater treatment by advanced oxidative process. *Sci. For.*, 45, 363, 2017.
28. Casadidio, C. *et al.*, Daptomycin-loaded biodegradable thermosensitive hydrogels enhance drug stability and foster bactericidal activity against staphylococcus aureus. *Eur. J. Pharm. Biopharm.*, 130, 260, 2018.
29. Othman, S.H., Bio-nanocomposite materials for food packaging applications: Types of biopolymer and nano-sized filler. *Agric. Agric. Sci. Proc.*, 2, 296, 2014.
30. Younes, I. and Rinaudo, M., Chitin and chitosan preparation from marine sources. Structure, properties and applications. *Mar. Drugs*, 13, 1133, 2015.
31. Honarkar, H. *et al.*, Applications of biopolymersI: Chitosan. *Monatsh. Chem.*, 140, 1403, 2009.
32. Anitha, A. *et al.*, Chitin and chitosan in selected biomedical applications. *Prog. Polym. Sci.*, 39, 1644, 2014.
33. Lins, L.C. *et al.*, Composite PHB/chitosan microparticles obtained by spray drying: Effect of chitosan concentration and crosslinking agents on drug release. *J. Braz. Chem. Soc.*, 25, 1462, 2014.
34. Laranjeira, M.C.M. and Fávere, V.T., Quitosana: Biopolímero funcional com potencial industrial biomédico. *Quim. Nova*, 32, 672, 2009.

35. Silva, H.S.R.C., Santos, K.S.C.R., Ferreira, E.I., Quitosana: Derivados hidrossolúveis, aplicações farmacêuticas e avanços. *Quim. Nova*, 29, 776, 2006.
36. Hou, L. and Wu, P., Exploring the hydrogen-bond structures in sodium alginate through two-dimensional correlation infrared spectroscopy. *Carbohydr. Polym.*, 205, 420, 2019.
37. Ibánez, J.P. and Umetsu, Y., Potential of protonated alginate beads for heavy metals uptake. *Hydrometallurgy*, 64, 89, 2002.
38. da Silva, T.L. *et al.*, Evaluation of incorporation of diclofenac sodium in dried sericin-alginate particles prepared by ionic gelation technique. *Chem. Eng. Trans.*, 43, 829, 2015.
39. Li, K. *et al.*, Preparation of chitosan-sodium alginate films through layer-by-layer assembly and ferulic acid crosslinking: Film properties, characterization, and formation mechanism. *Int. J. Biol. Macromol.*, 122, 485, 2019.
40. Ngo, T.M.P. *et al.*, Effects of zinc oxide nanoparticles on the properties of pectin/alginate edible films. *Int. J. Polym. Sci.*, 1, 1, 2018.
41. Cesur, S. *et al.*, Preparation and characterization of electrospun polylactic acid/sodium alginate/orange oyster shell composite nanofiber for biomedical application. *J. Aust. Ceram. Soc.*, 1, 1, 2019.
42. Bekin, S. *et al.*, Synthesis, characterization and bending behavior of electroresponsive sodium alginate/poly(acrylic acid) interpenetrating network films under an electric field stimulus. *Sens. Actuators B-Chem.*, 202, 878, 2014.
43. Sellimi, S. *et al.*, Structural, physicochemical and antioxidant properties of sodium alginate isolated from a Tunisian brown seaweed. *Int. J. Biol. Macromol.*, 72, 1358, 2015.
44. Ruvinov, E. and Cohen, S., Alginate biomaterial for the treatment of myocardial infarction: Progress, translational strategies, and clinical outlook: From ocean algae to patient bedside. *Adv. Drug Deliv. Rev.*, 96, 54, 2016.
45. Aguero, L. *et al.*, Alginate microparticles as oral colon drug delivery device: A review. *Carbohydr. Polym.*, 168, 32, 2017.
46. Dehghani, S., Hosseini, S.V., Regenstein, J.M., Edible films and coatings in seafood preservation: A review. *Food Chem.*, 240, 505, 2018.
47. Laurienzo, P., Marine polysaccharides in pharmaceutical applications: An overview. *Mar. Drugs*, 8, 2435, 2010.
48. Lee, K.Y. and Mooney, D.J., Alginate: Properties and biomedical applications. *Prog. Polym. Sci.*, 37, 106, 2012.
49. Mogoşanu, G.D. and Grumezescu, A.M., Natural and synthetic polymers for wounds and burns dressing. *Int. J. Pharm.*, 463, 127, 2014.
50. Wasupalli, G.K. and Verma, D., Molecular interactions in self-assembled nano-structures of chitosan-sodium alginate based polyelectrolyte complexes. *Int. J. Biol. Macromol.*, 114, 10, 2018.
51. Momoh, F.U. *et al.*, Development and functional characterization of alginate dressing as potential protein delivery system for wound healing. *Int. J. Biol. Macromol.*, 81, 137, 2015.

52. Zahran, M.K., Ahmed, H.B., El-Rafie, M.H., Surface modification of cotton fabrics for antibacterial application by coating with AgNPs–alginate composite. *Carbohydr. Polym.*, 108, 145, 2014.
53. Carvalho, A.M. *et al.*, Teores de hemicelulose, celulose e lignina em plantas de cobertura com potencial para sistema plantio direto no Cerrado in: *Boletim de Pesquisa e Desenvolvimento*, Embrapa Cerrados, edição 210, Planaltina city - District Federal (DF)–Brazil, 2010, http://www.cpac.embrapa.br/quadro/23.
54. Wild, P., Reith, H., Heeres, H., Biomass pyrolysis for chemicals. *Biofuels*, 2, 185, 2011.
55. Doherty, W.O.S., Mousavioun, P., Fellows, C.M., Value-adding to cellulosic ethanol: Lignin polymers. *Ind. Crops Prod.*, 33, 259, 2011.
56. Feldman, D., Lignin nanocomposites. *J. Macromol. Sci. A*, 53, 382, 2016.
57. Calvo-Flores, F.G. and Dobado, J.A., Lignin as renewable raw material. *ChemSusChem*, 3, 1227, 2010.
58. Figueiredo, P. *et al.*, Properties and chemical modifications of lignin: Towards lignin-based nanomaterials for biomedical applications. *Prog. Mater. Sci.*, 93, 233, 2018.
59. Constant, S. *et al.*, New insights into the structure and composition of technical lignins: A comparative characterization study. *Green Chem.*, 18, 2651, 2016.
60. Norgren, M. and Edlund, H., Lignin: Recent advances and emerging applications. *Curr. Opin. Colloid Interface Sci.*, 19, 409, 2014.
61. Azadi, P. *et al.*, Liquid fuels, hydrogen and chemicals from lignin: A critical review. *Renew. Sust. Energ. Rev.*, 21, 506, 2013.
62. Duval, A. and Lawoko, M., A review on lignin-based polymeric, micro- and-nano-structured materials. *React. Funct. Polym.*, 85, 78, 2014.
63. Stewart, D., Lignin as a base material for materials applications: Chemistry, application and economics. *Ind. Crops Prod.*, 27, 202, 2008.
64. Laurichesse, S. and Avérous, L., Chemical modification of lignins: Towards biobased polymers. *Prog. Polym. Sci.*, 39, 1266, 2014.
65. Munim, S.A. and Raza, Z.A., Poly(lactic acid) based hydrogels: Formation, characteristics and biomedical applications. *J. Porous Mater.*, 26, 881, 2018.
66. Holm, M.S., Saravanamurugan, S., Taarning, E., Conversion of sugars to lactic acid derivatives using heterogeneous zeotype catalysts. *Science*, 328, 602, 2010.
67. Nofar, M., Salehiyan, R., Ray, S.S., Rheology of poly (lactic acid)-based systems. *Polym. Rev.*, 59, 465, 2019.
68. Bohlmann, G.M., General characteristics, processability, industrial applications and market evolution of biodegradable polymers, in: *Handbook of Biodegradable Polymers*, C. Bastioli (Ed.), pp. 183–218, Rapra Technology limited, United Kingdom, 2005.
69. Souza, P.M.S. *et al.*, PLA and montmorilonite nanocomposites: Properties, biodegradation and potential toxicity. *J. Polym. Environ.*, 21, 738, 2013.

70. Arrieta, M.P. *et al.*, Disintegrability under composting conditions of plasticized PLA–PHB blends. *Polym. Degrad. Stab.*, 108, 307, 2014.
71. Weng, Y.X. *et al.*, Biodegration behavior of poly(butylene adipate-co-terephthalate) (PBAT), poly(lactic acid) (PLA), and their blend under soil conditions. *Polym. Test.*, 32, 918, 2013.
72. Kale, G. *et al.*, Biodegradability of polylactide bottles in real and simulated composting conditions. *Polym. Test.*, 26, 1049, 2007.
73. Ohkita, T. and Lee, S.H., Thermal degradation and biodegradability of poly-(lacticctic acid)/corn starch biocomposites. *J. Appl. Polym. Sci.*, 100, 3009, 2006.
74. Casasola, R. *et al.*, Electrospun poly lactic acid (PLA) fibres: Effect of different solvent systems on fibre morphology and diameter. *Polymer*, 55, 4728, 2014.
75. Oh, J.K., Polylactide (PLA)-based amphiphilic block copolymers: Synthesis, self-assembly, and biomedical applications. *Soft Matter*, 7, 5096, 2011.
76. Pacheco, D.P. *et al.*, Design of multifunctional polysaccharides for biomedical applications: A critical review. *Curr. Org. Chem.*, 22, 1222, 2018.
77. Peltzer, M. *et al.*, β-glucan, a promising polysaccharide for bio-based films developments for food contact materials and medical applications. *Curr. Org. Chem.*, 22, 1249, 2018.
78. Zhu, Y., Li, Z., Chen, J., Applications of lignin-derived catalysts for green synthesis. *Green Energy Environ.*, 4, 210, 2019.
79. Pulieri, E. *et al.*, Chitosan/gelatin blends for biomedical applications. *J. Biomed. Mater. Res.: Part A*, 86, 311, 2008.
80. Upadhyaya, L. *et al.*, Biomedical applications of carboxymethyl chitosans. *Carbohydr. Polym.*, 91, 452, 2013.
81. Liu, X.F. *et al.*, Antibacterial action of chitosan and carboxymethylated. *J. Appl. Polym. Sci.*, 79, 1324, 2001.
82. Liu, J. *et al.*, Preparation, characterization and antioxidant activity of phenolic acids grafted carboxymethyl chitosan. *Int. J. Biol. Macromol.*, 62, 85, 2013.
83. Chen, W. *et al.*, Synthesis and antioxidant properties of chitosan and carboxymethyl chitosan-stabilized selenium nanoparticles. *Carbohydr. Polym.*, 132, 574, 2015.
84. Narayanan, D., Jayakumar, R., Chennazhi, K.P., Versatile carboxymethyl chitin and chitosan nanomaterials: A review. *Wires Nanomed. Nanobiotechnol.*, 6, 574, 2014.
85. Kim, J.J. *et al.*, Immunoadjuvant efficacy of N-carboxymethyl chitosan for vaccination via dendritic cell activation. *J. Med. Food*, 17, 268, 2014.
86. Prabaharan, M., Chitosan-based nanoparticles for tumor-targeted drug delivery. *Int. J. Biol. Macromol.*, 72, 1313, 2015.
87. Shaima, C., Moorthi, P.V., Kutty, S.N., *In vitro* anticancer activity of 5' fluorouracil coated chitosan nanoparticle. *Int. J. Curr. Pharm. Res.*, 8, 6, 2016.

88. Zhang, X. *et al.*, A pH-sensitive nanosystem based on carboxymethyl chitosan for tumor-targeted delivery of daunorubicin. *J. Biomed. Nanotechnol.*, 12, 1688, 2016.
89. Yu, S. *et al.*, A novel pH-induced thermosensitive hydrogel composed of carboxymethyl chitosan and poloxamer cross-linked by glutaraldehyde for ophthalmic drug delivery. *Carbohydr. Polym.*, 155, 208, 2017.
90. Upadhyaya, L. *et al.*, The implications of recent advances in carboxymethyl chitosan based targeted drug delivery and tissue engineering applications. *J. Control. Release*, 186, 54, 2014.
91. Croisier, F. and Jérôme, C., Chitosan-based biomaterials for tissue engineering. *Eur. Polym. J.*, 49, 780, 2013.
92. Requena, R. *et al.*, Biopolymers carrying essential oils, or their compounds, for food antimicrobial packaging. *Curr. Org. Chem.*, 22, 1141, 2018.
93. Wei, S. *et al.*, Preparation and thermal performances of microencapsulated phase change materials with a nano-Al_2O_3-doped shell. *J. Therm. Anal. Calorim.*, 138, 233–241, 2019.
94. Auras, R., Harte, B., Selke, S., An overview of polylactides as packaging materials. *Macromol. Biosci.*, 4, 835, 2004.
95. Tsuji, H. and Ikada, Y., Crystallization from the melt of poly(lactide)s with different optical purities and their blends. *Macromol. Chem. Phys.*, 197, 3483, 1996.
96. Singh, S.K., Solubility of lignin and chitin in ionic liquids and their biomedical applications. *Int. J. Biol. Macromol.*, 132, 265, 2019.
97. Pillai, C.K.S., Paul, W., Sharma, C.P., Chitin and chitosan polymers: Chemistry, solubility and fiber formation. *Prog. Polym. Sci.*, 34, 641, 2009.
98. Rinaudo, M., Main properties and current applications of some polysaccharides as biomaterials. *Polym. Int.*, 57, 397, 2008.
99. Zare, E.N., Makvandi, P., Tay, F.R., Recent progress in the industrial and biomedical applications of tragacanth gum: A review. *Carbohydr. Polym.*, 212, 450, 2019.
100. Moghaddama, R.H. *et al.*, Synthesis of composite hydrogel of glutamic acid, gum tragacanth, and anionic polyacrylamide by electron beam irradiation for uranium (VI) removal from aqueous samples: Equilibrium, kinetics, and thermodynamic studies. *Carbohydr. Polym.*, 206, 352, 2019.
101. Markert, C.D. *et al.*, Characterizing the micro-scale elastic modulus of hydrogels for use in regenerative medicine. *J. Mech. Behav. Biomed. Mater.*, 27, 115, 2013.
102. Chang, Y.R. *et al.*, Automated AFM force curve analysis for determining elastic modulus of biomaterials and biological samples. *J. Mech. Behav. Biomed. Mater.*, 37, 209, 2014.
103. Armentano, I. *et al.*, Nanostructured biopolymer-based materials for regenerative medicine applications. *Curr. Org. Chem.*, 22, 1193, 2018.

104. Chen, H. *et al.*, Preparation, characterization, and properties of chitosan films with cinnamaldehyde nanoemulsions. *Food Hydrocolloids*, 61, 662, 2016.
105. Raphey, V.R. *et al.*, Advanced biomedical applications of carbon nanotube. *Mater. Sci. Eng. C*, 100, 616, 2019.
106. Anitha, A. *et al.*, Development of mucoadhesive thiolated chitosan nanoparticles for biomedical applications. *Carbohydr. Polym.*, 83, 66, 2011.
107. Rao, K. *et al.*, Gum tragacanth stabilized green gold nanoparticles as cargos for Naringin loading: A morphological investigation through AFM. *Carbohydr. Polym.*, 174, 243, 2017.
108. Siqueira, P. *et al.*, Three-dimensional stable alginate-nanocellulose gels for biomedical applications: Towards tunable mechanical properties and cell growing. *Nanomaterials*, 9, 1, 2019.
109. Sehaqui, H., Zhou, Q., Berglund, L.A., High-porosity aerogels of high specific surface area prepared from nanofibrillated cellulose (NFC). *Compos. Sci. Technol.*, 71, 1593, 2011.
110. Park, J.W. *et al.*, Biodegradable polymer blends of poly(lactic acid) and starch. *Korea Polym. J.*, 7, 93, 1999.
111. Vasile, C. *et al.*, Biocompatible materials based on plasticized poly(lactic acid), chitosan and rosemary ethanolic extract I. Effect of chitosan on the properties of plasticized poly(lactic acid) materials. *Polymers*, 1, 941, 2019.
112. Larrañeta, E. *et al.*, Synthesis and characterization of lignin hydrogels for potential applications as drug eluting antimicrobial coatings for medical materials. *ACS Sustainable Chem. Eng.*, 6, 9037, 2018.
113. Rajalekshmy, G.P. *et al.*, An overview on the potential biomedical applications of polysaccharides, in: *Functional polysaccharides for biomedical applications*, S. Maiti and S. Jana (Eds.), pp. 33–99, Elsevier Ltda, Woodhead Publishing, 2019.
114. Kalantaria, K. *et al.*, Biomedical applications of chitosan electrospun nanofibers as a green polymer-Review. *Carbohydr. Polym.*, 207, 588, 2019.
115. Cruz, G. *et al.*, Investigation into the physical-chemical properties of chemically pretreated sugarcane bagasse. *J. Therm. Anal. Calorim.*, 132, 1039, 2018.
116. Viel, M. *et al.*, Hemp-straw composites: Gluing study and multi-physical characterizations. *Materials*, 12, 1, 2019.
117. Saltzman, W.M. and Kyriakides, T.R., Cell interactions with polymers, in: *Principles of Tissue Engineering*, R.P. Lanza, R. Langer, W.L. Chick (Eds.), pp. 385–406, Academic Press, Georgetown, 2014.
118. Kister, G. and Cassanas, G., Vert, M., Effects of morphology, conformation and configuration on the IR and Raman spectra of various poly(lactic acid)s. *Polymer*, 39, 267, 1998.
119. Ferrández-Montero, A. *et al.*, Development of biocompatible and fully bioabsorbable PLA/Mg films for tissue regeneration applications. *Acta Biomater.*, 98, 114–124, 2019.

120. Han, J. *et al.*, Alginate-chitosan/hydroxiapatite polyelectrolyte complex porous scaffolds: Preparation and characterization. *Int. J. Biol. Macromol.*, 46, 199, 2009.
121. Koo, B.W. *et al.*, Organosolv pretreatment of *Liriodendron tulipifera* and simultaneous saccharification and fermentation for bioethanol production. *Biomass Bioenergy*, 35, 1833, 2011.
122. Du, J. *et al.*, Understanding the toxicity of carbon nanotubes in the environment is crucial to the control of nanomaterials in producing and processing and the assessment of health risk for human: A review. *Environ. Toxicol. Pharmacol.*, 36, 451, 2013.
123. Fu, P.P. *et al.*, Mechanism of nanotoxicity: Generation of reactive oxygen species. *J. Food Drug Anal.*, 22, 64, 2014.
124. Zakaria, Z. *et al.*, Mechanical properties and morphological characterization of PLA/chitosan/epoxidized natural rubber composites. *Adv. Mater. Sci. Eng.*, 2013, 1–7, 2013.
125. Bijarimi, M. *et al.*, Poly(lactic acid)/poly(ethylene glycol) blends: Mechanical, thermal and morphological properties. *AIP Conf. Proc.*, 1727, 020002, 2016.
126. Arrieta, M.P. *et al.*, Effect of chitosan and catechin addition on the structural, thermal, mechanical and disintegration properties of plasticized electrospun PLA-PHB biocomposites. *Polym. Degrad. Stab.*, 132, 145, 2016.
127. Wróblewska-Krepsztul, J. *et al.*, Biopolymers for biomedical and pharmaceutical applications: Recent advances and overview of alginate electrospinning. *Nanomaterials*, 9, 7, 2019.
128. Zhang, X. *et al.*, Marine biomaterial-based bioinks for generating 3D printed tissue constructs. *Mar. Drugs*, 16, 12, 2018.
129. Seo, Y. *et al.*, Electrostatically interactive injectable hydrogels for drug delivery. *Tissue Eng. Regen. Med.*, 15, 513, 2018.
130. Soorbaghi, F.P. *et al.*, Bioaerogels: Synthesis approaches, cellular uptake, and the biomedical applications. *Biomed. Pharmacother.*, 111, 964, 2019.
131. Salesa, B. *et al.*, Carbon nanofibers in pure form and in calcium alginate composites films: New cost-effective antibacterial biomaterials against the life-threatening multidrug-resistant Staphylococcus epidermidis. *Polymers*, 11, 1, 2019.
132. Noh, I. *et al.*, 3D printable hyaluronic acid-based hydrogel for its potential application as a bioink in tissue engineering. *Biomater. Res.*, 23, 1, 2019.
133. Grumezescu, V. *et al.*, Nanomagnetite-embedded PLGA spheres for multipurpose medical applications. *Materials (Basel)*, 12, 2521, 2019.
134. Kravanja, G. *et al.*, Chitosan-based (Nano)materials for novel biomedical applications. *Molecules*, 24, 1960, 2019.
135. Chen, Y. *et al.*, Highly mechanical properties nanocomposite hydrogels with biorenewable lignin nanoparticles. *Int. J. Biol. Macromol.*, 128, 414, 2019.
136. Lee, B.K., Yun, Y., Park, K., PLA micro- and nano-particles. *Adv. Drug Deliv. Rev.*, 107, 176, 2016.

137. Giammona, G. and Craparo, E.F., Biomedical applicationsof polylactide (PLA) and its copolymers. *Molecules*, 23, 980, 2018.
138. Contreras-Cáceres, R. *et al.*, Electrospun nanofibers: Recent applications in drug delivery and cancer therapy. *Nanomaterials*, 9, 656, 2019.
139. Eisenbrey, J.R. *et al.*, Development and optimization of a doxorubicin loaded poly(lactic acid) contrast agent for ultrasound directed drug delivery. *J. Control. Release*, 143, 38, 2010.
140. Li, J. *et al.*, Novel strategy to fabricate PLA/Au nanocomposites as an efficient drug carrier for human Leukemia cells *in Vitro*. *Nanoscale Res. Lett.*, 6, 1, 2011.
141. Xie, Z. *et al.*, Immune cell-mediated biodegradable theranostic nanoparticles for melanoma targeting and drug delivery. *Small*, 13, 1603121, 2017.
142. Xin, X., Hussain, M., Mao, J.J., Continuing differentiation of human mesenchymal stem cells and induced chondrogenic and osteogeniclineages in electrospun PLGA nanofiber scaffold. *Biomaterials*, 28, 316, 2007.
143. Narayanan, G. *et al.*, Poly (lacticacid)-based biomaterials for orthopaedicre generative engineering. *Adv. Drug Deliv. Rev.*, 107, 247, 2016.
144. Bayer, I.S., Thermomechanical properties of Polylactic acid-graphene composites: A state-of-the-art review for biomedical applications. *Materials*, 1, 748, 2017.

5
Waterborne Adhesives

Sukanya Pradhan

Department of Applied Chemistry, Centurion University of Technology and Management, Odisha, India

Abstract
There are alarming concerns over the materials of conventional solvent-based adhesives which are not considered environmentally benign. The utilization and promotion of water-based adhesives has become a new trend in order to minimize the environmental concerns. Green materials are gaining more attention in all fields owing to their improved environmental footprint and due to the null dependence from petroleum resources. This holds true in the field of waterborne adhesives, where the motive is to check VOC emission and waste water creation by using water as solvent/co-solvent. However, their sustainability has not been proved enough for the commercial use as replacement of conventional adhesives. The aim of this chapter is therefore to highlight the performance characteristics of waterborne adhesive and its impact on the environment.

Keywords: Waterborne, adhesive, VOC, polyurethane, epoxy

5.1 Introduction

Adhesives are those materials which are capable of holding different substrates that meets the demands of various applications ranging from industry to medical use [1, 2]. The environmental protection as well as high cost of oil derivatives and legislative pressure have attracted the scientific community to conduct research on the development of sustainable adhesives with no use of solvents. The emergence of renewable natural resources-based waterborne adhesive has replaced or reduced the production of synthetic adhesive by reducing VOC emission. Waterborne materials can be fabricated

Email: pradhan.sukanya2@gmail.com

Inamuddin, Rajender Boddula, Mohd Imran Ahamed and Abdullah M. Asiri (eds.) Green Adhesives: Preparation, Properties and Applications, (121–128) © 2020 Scrivener Publishing LLC

into coatings and adhesives, containing less or no co-solvent, that form films at ambient temperature. The hydrophilicity of these polymeric materials is modified into water soluble adhesives by the incorporation of polar or hydrophilic segments into the polymer backbone [3]. These materials exhibit excellent adhesion to many substrates ranging from glass to polymeric fibers. These eco-friendly polymers are non-toxic, non-flammable, and they do not possess any negative impact on environment [4].

5.1.1 Motivation for the Use of Waterborne Adhesives

This section presents following incentives that have encouraged the production of waterborne adhesive system.

5.1.1.1 Sustainability and Environment Regulations

In the present scenario, there has been a tremendous demand in the utilization of renewable resources in order to improve the sustainability of the polymeric products. This positive drive towards sustainability is due to the public awareness about the environment and the stringent regulations as well as the threat associated with the petroleum price volatility. Additionally, this has also forced the adhesive industry to produce waterborne adhesives with high solid content as substitutes for solvent-based adhesives [5].

5.1.1.2 Circular Economy

The formation of circular economy can be induced by producing adhesives derived from renewable resources. Under this technique, the bioderived materials will definitely reduces the carbon footprints with high biodegradability and less or no toxicity characteristics as compared to petro-based solventborne adhesives [6].

5.1.1.3 Avoid Harmful Emissions

This is the primary factor that has accelerated the use of water as solvent instead of destructive substitutes in order to check the harmful emissions during its processing as well as during the service life [7].

5.1.1.4 Development of Novel and Sustainable End Products

Another incentive for the production of waterborne adhesives is the investment in the adhesive and coatings market for the development of novel and sustainable products. This highly competitive market demands the generation of customized products with desirable characteristics to enable distinction and increase loyalty [8, 9].

5.1.2 Environmental Effects and Mankind Toxicity Analysis

The completely cured adhesives are considered to be non-toxic and safe [8]. However, they may produce hazardous emissions during processing that imposes negative impact on the environment as well as to the mankind. It also releases volatile organic compounds and other waste materials depending on the type of solvents used. For example, toluene which is used as diluent in conventional thermal-melt adhesives emits VOCs and hence considered as environmental damaging solvent [10]. Figure 5.1 represents the legally permissible exposure limits (PEL). Various researchers have proved that for the entire life cycle, the solventborne adhesives have greater negative environmental impact than the waterborne adhesives [9, 11, 12].

The solventborne adhesives also possess risk of skin irritation, eye sight damage due to the exposure of the volatile components emitted by the solvent. Additionally, this category of adhesive systems also indirectly affects the central nervous system by penetration through the skin. The above discussed shortcomings of solvent-based adhesives can be overcome by the utilization of water as solvent. In a report, bacterial polysaccharide FucoPol derived from the bacterium Enterobacter A47 was synthesized and its solution with the assistance of water has been proved to be a great potential adhesive for wood-wood, glass-glass, cellulose acetate-cellulose acetate, and cardboard-cardboard joints with excellent shear strength value for wood bonding [13]. However, its use is restricted to limited applications

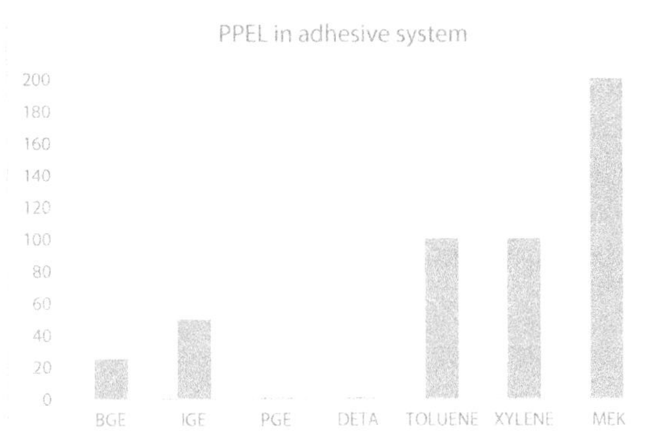

Figure 5.1 PEL in adhesive systems.

but its low quantity requirement and the fact of using water as a solvent signifies its greatest compatible characteristics with the environment.

5.2 Performance of Waterborne Adhesives: An Overview

5.2.1 Waterborne Polyurethane (WBPU) Adhesives

Polyurethanes are composed of organic units combined together by urethane linkages and these macromolecules are obtained as a product by the reaction of monomeric diisocyanate or polymeric isocyanate with the assistance of a hydroxyl functionalities containing monomer in presence of appropriate catalyst and certain additives [14]. Polyurethanes are generally hydrophobic in nature and hence not compatible with water. There are certain modifications such as incorporation of ionic groups (quaternary ammonium, carboxylate, or sulfonate groups), polar groups or non-ionic hydrophilic segments into the polymeric backbone to lower down the hydrophobicity thereby making it water soluble [15]. The first aqueous polyurethane dispersion was introduced in the late 1960s from which the idea of producing water-based polyurethanes with more hydrophilicity is being inherited [16]. Waterborne polyurethane finds applications in a wide range such as as coatings and adhesives, ink binding, dyes for textile industry, glass fibers, paper, films packaging, additives, sealants, thickeners, etc.

5.2.1.1 Chemical Structure of Waterborne PU

WBPU consists of two segments—the primary polymeric backbone and the emulsifier which are hydrophillic in nature. This emulsifier helps in the formation of WBPU dispersions in which the PU particles forms the discontinuous phase and the liquid in which it is dispersed hold the continuous phase [17]. The colloidal WBPU dispersion is stable in a wide range of pH levels and it depends upon the coalescence and total surface area reduction. The three types of hydrophilic groups which are employed for the synthesis of ionomers are cationic, anionic, and non-ionic WBPU dispersions.

5.2.1.2 Performances of WBPU Adhesives

The use of WBPU methodology is dominated in both academic and industrial realm. Various researchers have been reported the successful synthesis of biobased monomers to formulate WBPU dispersions. The overall performance characteristics could be further improved by the incorporation of renewable resource fillers, for instance, tannins and lignins. The rigid structural units of lignin permits its use as a cheap reinforcement as fillers. However, exploring of lignin in WBPU coating and adhesive materials is very rare. The major shortcoming of WBPU/lignin dispersion is its poor stability and storage stability period. This has been overcome by a study by Liu *et al.* [18] in which modified lignin by the addition of diethylenetriamine through mannic reaction is being used.

Several efforts have been taken to prepare formaldehyde free adhesives from soybean flour due to its wide availability and cost-effective nature as an alternative to formaldehyde-based adhesives [19–21]. But, due to its poor water resistance performance, it is limited to certain specific applications. Many research works have been performed to increase the water resistance by undergoing certain modifications such as exposing hydrophobic subunits and then employing methods such as enzymatic modification, blending, and crosslinking, etc. [22, 23]. In a study, soybean flour-based adhesive with excellent water resistance has been reported which has exhibited a significant wet shear strength by the incorporation of WBPU [24].

5.2.2 Waterborne Epoxy Adhesive

During the last several decades, formaldehyde resins such as urea-formaldehyde (UF) and phenol-formaldehyde (PF) have dominated the adhesive industry owing to its low cost, low viscosity, and aqueous nature. But, the major drawback lies with its carcinogenic characteristics. Epoxy adhesives have come out to fulfill all the desired requirements of the adhesive industry. Besides its excellent wettability characteristics on cellulosic substrates, they could not be used as wood composite adhesive due to highly expensive, use of organic solvents, and highly viscous properties [25]. These drawbacks were supposed to be overcome by the emergence of waterborne epoxy adhesives as they possess low viscosity and good wetting and penetration properties which are comparable to that of UF and PF resins. However, it could not be proved as a potential candidate for UF and MF alternatives due to the usage of petroleum-based waterborne polyamidoamine or aromatic polycarboxylic acids which in turn increases the overall costs [26]. Hence, the utilization of renewable resource-based waterborne epoxy and curing agent has become a great interest for the scientific community for the synthesis of waterborne epoxy.

5.3 Conclusions

In the current scenario, waterborne adhesives derived from bioresources with water as solvent have been highlighted due to greener approach, cost effective, wide availability, biodegradability, and a better potential candidate as end use applications. The rapid hike in solvent price, increased environmental concerns, and ease of cleaning up the reactor system are some of the issues that have driven the use of water as solvent during the synthesis of waterborne adhesive systems. It is worth mentioning that the waterborne adhesives possess poor mechanical, thermal instability, and weak hydrolytic strength in wet conditions which can be resolved by the incorporation of nanoparticles in the near future.

References

1. Orgilés-Calpena, E., Arán-Aís, F., Torró-Palau, A.M., Montiel-Parreño, E., Orgilés-Barceló, C., Synthesis of polyurethanes from CO2-based polyols: A challenge for sustainable adhesives. *Int. J. Adhes. Adhes.*, 67, 63–68, 2016.
2. Cornille, A., Michaud, G., Simon, F., Fouquay, S., Auvergne, R., Boutevin, B., Caillol, S., Promising mechanical and adhesive properties of isocyanate-free poly (hydroxyurethane). *Eur. Polym. J.*, 84, 404–420, 2016.
3. Pradhan, S., Mohanty, S., Nayak, S.K., Waterborne epoxy adhesive derived from epoxidized soybean oil and dextrin: Synthesis and characterization. *Int. J. Polym. Anal. Charact.*, 22, 4, 318–329, 2017.
4. Khanlari, S. and Dubé, M.A., Bioadhesives: A review. *Macromol. React. Eng.*, 7, 11, 573–587, 2013.
5. Heinrich, L.A., Future opportunities for bio-based adhesives–advantages beyond renewability. *Green Chem.*, 21, 8, 1866–1888, 2019.
6. Gómez, E.F. and Michel, F.C., Jr., Biodegradability of conventional and bio-based plastics and natural fiber composites during composting, anaerobic digestion and long-term soil incubation. *Polym. Degrad. Stab.*, 98, 12, 2583–2591, 2013.
7. Väisänen, A.J., Hyttinen, M., Ylönen, S., Alonen, L., Occupational exposure to gaseous and particulate contaminants originating from additive manufacturing of liquid, powdered, and filament plastic materials and related post-processes. *J. Occup. Environ. Hyg.*, 16, 3, 258–271, 2019.
8. Gadhave, R.V., Mahanwar, P.A., Gadekar, P.T., Bio-renewable sources for synthesis of eco-friendly polyurethane adhesives, *Open Journal of Polymer Chemistry*, 7, 57–75, 2017.
9. Yang, M. and Rosentrater, K., Environmental effects and economic analysis of adhesives: A review of life cycle assessment (LCA) and techno-economic

analysis (TEA), in: *2015 ASABE Annual International Meeting*, American Society of Agricultural and Biological Engineers, p. 1, 2015.
10. De Gray, D., PUR adhesives offer solutions for assembly challenges. *Adhes. Age*, 41, 5, 23–24, 1998.
11. Noreen, A., Zia, K.M., Zuber, M., Tabasum, S., Zahoor, A.F., Bio-based polyurethane: An efficient and environment friendly coating systems: A review. *Prog. Org. Coat.*, 91, 25–32, 2016.
12. Ionescu, M., Petrović, Z.S., Wan, X., Ethoxylated soybean polyols for polyurethanes. *J. Polym. Environ.*, 15, 4, 237–243, 2007.
13. Araújo, D., Alves, V.D., Campos, J., Coelhoso, I., Sevrin, C., Grandfils, C., Reis, M.A., Assessment of the adhesive properties of the bacterial polysaccharide FucoPol. *Int. J. Biol. Macromol.*, 92, 383–389, 2016.
14. Panda, S.S., Panda, B.P., Nayak, S.K., Mohanty, S., A review on waterborne thermosetting polyurethane coatings based on castor oil: Synthesis, characterization, and application. *Polym.-Plast. Technol. Eng.*, 57, 6, 500–522, 2018.
15. Honarkar, H., Waterborne polyurethanes: A review. *J. Dispersion Sci. Technol.*, 39, 4, 507–516, 2018.
16. Tao, C., Bao, J., Cheng, Q., Huang, Y., Xu, G., Preparation of waterborne polyurethane adhesives based on macromolecular-diols containing different diisocyanate. *J. Adhes.*, 95, 9, 814–833, 2019.
17. Minnath, M.A. and Purushothaman, E., Blends and Interpenetrating Polymer Networks Based on Polyurethane Polymers With Natural and Synthetic Rubbers, in: *Polyurethane Polymers*, pp. 17–44, Elsevier, Amsterdam, Netherland, 2017.
18. Liu, J., Liu, H.F., Deng, L., Liao, B., Guo, Q.X., Improving aging resistance and mechanical properties of waterborne polyurethanes modified by lignin amines. *J. Appl. Polym. Sci.*, 130, 3, 1736–1742, 2013.
19. Mai, C., Kües, U., Militz, H., Biotechnology in the wood industry. *Appl. Microbiol. Biotechnol.*, 63, 5, 477–494, 2004.
20. Mancera, C., El Mansouri, N.E., Pelach, M.A., Francesc, F., Salvadó, J., Feasibility of incorporating treated lignins in fiberboards made from agricultural waste. *Waste Manage.*, 32, 10, 1962–1967, 2012.
21. Rahman, M., Zahir, M., Kim, H., Synthesis and properties of waterborne polyurethane (wbpu)/modified lignin amine (mla) adhesive: A promising adhesive material. *Polymers*, 8, 9, 318, 2016.
22. Kim, B.S. and Kim, B.K., Enhancement of hydrolytic stability and adhesion of waterborne polyurethanes. *J. Appl. Polym. Sci.*, 97, 5, 1961–1969, 2005.
23. Zhang, B.H., Fan, B., Li, M., Zhang, Y.H., Gao, Z.H., Effects of thermal treatment on the properties of defatted soya bean flour and its adhesion to plywood. *R. Soc. Open Sci.*, 5, 5, 180015, 2018.
24. Wang, Y., Deng, L., Fan, Y., Preparation of Soy-Based Adhesive Enhanced by Waterborne Polyurethane: Optimization by Response Surface Methodology. *Adv. Mater. Sci. Eng.*, 2018, 1–8, 2018.

25. Badr, M.M., Amer, A.A., Shehat, A.S., Synthesis and characterization of waterborne epoxy resins for coating application. *Aust. J. Basic Appl. Sci.*, 4, 6, 1376–1382, 2010.
26. Liu, X., Wang, Y., Cao, Y., Yadama, V., Xian, M., Zhang, J., Study of dextrin-derived curing agent for waterborne epoxy adhesive. *Carbohydr. Polym.*, 83, 3, 1180–1184, 2011.

6

Using Polyfurfuryl Alcohol as Thermoset Adhesive/Sealant

Rakesh Kumar[1]* and Rajnish Kumar[2]

[1]*Department of Biotechnology, Central University of South Bihar, Gaya, India*
[2]*Department of Mathematics, Birla Institute of Technology Mesra, Patna Campus, India*

Abstract
Furfuryl alcohol (FA) can be used as an adhesive and sealant through temperature dependent controlled polymerization in presence of an acid catalyst. The difficulties are encountered in the conversion of FA into polyfurfuryl alcohol (PFA) leading to the exothermic reaction often leading to dangerous explosion. The commercial development of this processes/technology requires a specification of the conditions for the curing of biopolymer sealant with definite thickness and surface area and also to monitor blistering phenomena during curing reactions to get PFA sealant. In this book chapter, we have explored FA as an adhesive and PFA as a sealant. Also, a small experiment has been performed to depict the challenges being faced while curing of PFA sealant with respect to thickness (5 to 10mm), temperature (50°C to 60°C) and the concentration of acid catalyst (0.3 to 0.5 phr). Experimental data were validated with the curve fitting models with help of MATLAB software package. The best fitted model was reported for each of the parameters studied in this book chapter.

Keywords: Furfuryl alcohol, acid catalyst, adhesive, sealant, temperature, thickness, mathematical validation, blistering

**Corresponding author*: krrakesh72@gmail.com; rakeshkr@cusb.ac.in

Inamuddin, Rajender Boddula, Mohd Imran Ahamed and Abdullah M. Asiri (eds.) Green Adhesives: Preparation, Properties and Applications, (129–144) © 2020 Scrivener Publishing LLC

6.1 Introduction

Sealants are commonly used to minimize unwanted gaps during construction process, which may results in seepage or may not look ethical. Sealants and adhesives are interrelated because sealants have also adhesive properties. These days, thermoset-based synthetic sealants and adhesives such as expoxies and phenolics are widely used for bonding as well as in dental application [1, 2].

Epoxy-based sealants posses' high strength, high toughness, and once applied it ensures effective bonding. In addition, epoxy-based sealants are chemically inert due to which it can easily resist environmental or chemical changes during application. Similar behavior, as observed for epoxy-based sealants, is shown by phenolic-based sealants that is composed of phenol and formaldehyde resin. Both epoxy and phenolic-based sealants cure at room temperature or even at elevated temperature [3]. Phenolic sealants can be available in powder, liquid, and film form. Thermoset polymers such as epoxies and phenolics exist as adhesives in the different form such as paste, liquid, film, and tape. Epoxy-based adhesives fall in the category of structural adhesives and can carry significant stresses that make it feasible in structural applications.

Thermosetting structural adhesives or sealants are normally available in two-part forms. They are mixed through carefully under controlled stoichiometric ratio. The product can be cured within the desired time window. Thermosetting structural adhesives or sealants are also available as one-part forms. In this category, the resin and hardener (cross-linking agent) are already mixed together. The challenge for one-part form-based sealants or adhesives is that it must be kept at sufficiently low temperature so that the reaction does not occur prematurely. Also, thermosetting sealants and adhesives falling in this category have limited shelf life and often must be stored at low temperatures.

The mechanism of working of thermoset-based adhesives and sealants are related to

(i) Temperature and curing time—The phenolic or epoxy-based sealants can be applied on the unwanted gaps. If the curing takes places at room/elevated temperature then the applied sealants are left undisturbed for desired time at room/elevated temperature. The presence of moisture in the air is responsible for curing of thermoset-based sealants.

(ii) Agents used for curing—Thermoset-based sealants are supplied in different packs. One pack contains resin and another pack contains hardner. Both the components are mixed together in specified compositions and are applied on the unwanted gaps. Dicyandiamide and polyamido-amine can be used as curing agent for epoxy bonded steel joints [4].

The above discussion relates to synthetic-based thermoset polymers as adhesive and sealant. Bio-based polymers can be biodegradable [e.g., poly-lactic acid (PLA), soy protein isolate (SPI) or starch] or non-degradable [e.g., polyfurfuryl alcohol (PFA)]. Biodegradable polymers with reproducible resource, good biocompatibility, and processabilty have a significant potential in food industry, agriculture, bioscience, and biotechnology [5, 6]. The limitations in using biopolymers such as PLA, SPI, and PFA as plastics have been explored by several authors [7–10]. Bio-based polymers are attracting increased attention as adhesives and sealants due to environmental concerns and the realization that global petroleum resources are finite [11]. Soy protein and lignin biopolymer have been explored extensively as a bioadhesives [12, 13]. Photo polymerized gelatin falling in the category of biopolymer has been used as elastic tissue dental sealant [14].

Furfuryl alcohol (FA) is obtained from renewable resources, i.e., sugarcane bagasse and is stated as potential green thermoset resin [15–17]. However, there are very few reports available for virgin polyfurfuryl alcohol (PFA)-based biopolymer sheet [8–10]. It has been well established that the presence of conjugated sequences in PFA give black colored cured material [9]. Researchers have encountered the difficulties in the conversion of FA as monomer into PFA. This is attributed to the exothermic reaction often leading to dangerous explosion, which can be overcome by the approaches presented in the South African Patent and World Patent filed by Kumar *et al.*, [18, 19]. The processes provided herein can be used to prepare various shapes and sizes of reinforced or unreinforced molded products that can also be used as a replacement for existing plastic sheets and composites in several sectors. It is important to monitor the curing reaction in the formation of PFA-based materials due to explosive nature of the reactions. Sometimes, slight change in parameter leads to blistering which can be also counted as explosion but at very low magnitude. So, monitoring curing reaction is one of the important aspects for studying PFA-based thermosetting resins as sealants or adhesives. Recently, there has been report showing the possible adhesion of aluminium wire with PFA [20].

Cure sensor research was initiated in early 1990s for thermoset resins [21]. Ultrasonic Dynamic Mechanical Analysis (UDMA) has been used for monitoring the physical changes taking place during curing and the potential for online monitoring during thermoset polymer [22].

To develop cost-effective polymer sheet, scientists have used numerical process modeling. It is an important tool in the manufacturing sector that can provide optimum temperature and other related parameters that are helpful during the curing of the polymeric materials [23, 24]. Kinetics models relating to curing behavior epoxy thermoset resin incorporated with soy protein has been reported [25]. In one research article, the validation of processing model was done by using experimental data from the bench scale unit [26].

6.2 Furfuryl Alcohol as Adhesives

Structurally furfural is an aldehyde of furan. In 2005, a global production of about 250×10^3 t/a was reported [27]. FA can be polymerized by the addition of a strong acid such as sulphuric acid, para-toulene sulphonic acid (PTSA), an alkane sulphonic acid having 2 to 14 carbon atoms [28]. FA is used as a foundry sand binder [29]. The polymerization of FA has been investigated in presence of spent-liquor and tannin extracts. FA-phenolic resins can be explored as combination of bio-synthetic polymer-based adhesive. At present in wood industry, formaldehyde-based adhesives, which are completely synthetic in nature, are widely used [30].

The furfuryl alcohol-spent liquor (FA-SL) resins have been explored as bioadhesives and it has been used to glue solid wood. The properties of bioadhesives were dependent on the composition of FA-SL formulations [31]. In the formulated FA-SL bioadhesives, the content of FA and SL were 20% and 80%, respectively. The FA-SL composition in presence of less than 2% sulphuric acid catalyst was exposed to 120°C that resulted in very weak bonding. On the other hand, strong bonding was observed when the FA and SL were mixed in equal proportion, i.e., 50% and 50%, respectively, and the solid woods were glued at 180°C instead of 120°C. The result obtained was comparable to that obtained for commercial polyvinylacetate adhesive used to glue solid wood.

FA-SL and FA-Taninin (FA-T) adhesives have been used to glue wood to produce solid panels. The appearance of FA-SL and FA-T boards are reported in Figure 6.1. The FA-T glue unlike FA-SL panels showed black adhesive spots and unglued wood chips. It has been reported that the lower reactivity between FA and SL helps the glue to distribute evenly over the

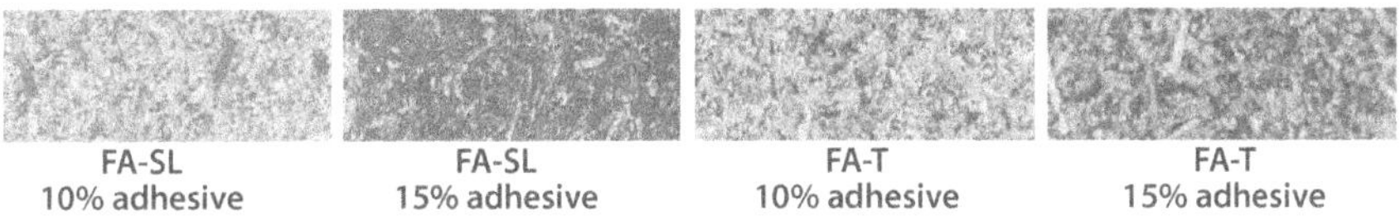

Figure 6.1 Photograph of fabricated particle board using bioadhesives (Reproduced From Reference 31, Open Access).

panel (Figure 6.1). Lower content (10%) of FA-SL and FA-T adhesives impart lower mechanical properties as compared to higher content (15%) of FA-SL and FA-T adhesives as well as FA-SL-T adhesives (mixed formulations). The mixed formulations showed internal bonds of ~0.5 MPa and elastic modulus of ~10 MPa. It has been reported by Abdullah and Pizza that FA-T-based bioadhesive has been used to prepare particleboard [32]. The particle board prepared from FA-SL compositions were not resistant to water because it showed a thickness swelling of around 100%.

Wang *et al.* have reported the oxidation of FA to furaldehyde over monolayer tiatanate nanosheet under visible light irradiation. Not only monomer of FA but oligomers of FA can also be prepared using tungsten oxide catalysts.

6.3 Polyfurfuryl Alcohol as Sealants

FA and PTSA monohydrate were purchased from Sigma-Aldrich and were used as received to explore PFA as sealants. The density of FA as mentioned in the manufacturer's data sheet is 1.135 gm/ml. Circular silicon-based mold of 10 cm diameter was used for studying PFA as sealants. The area of the silicon mold was 78.5 cm^2.

About 0.015 g of PTSA monohydrate was dissolved in 5 ml water to prepare catalyst solution. The catalyst solution was added in dropwise manner in FA at room temperature. Here, FA is a monomer from which PFA can be polymerized in presence of PTSA as an acid catalyst. Since the surface finish of the PFA-based material after polymerization is dependent of time of curing, thickness of the cured material, and the content of the catalyst so we have changed these three variables during curing of PFA.

In the first case, temperature and catalyst concentration were kept constant at 50±2°C and 0.3 parts per hundred (phr), respectively. The thickness of PFA sealants was varied in order to optimize the thickness. In the second case, temperature and thickness of PFA-based sealants were kept constant as 50°C and 5 mm and the contents of catalyst were varied. Lastly, at optimized thickness and catalyst concentration, temperature to cure PFA-based sealants was varied.

We have used curve fitting Toolbox in MATLAB software package to calculate the correlation coefficients for each of three different parameters and selected the best fitted curve for each of the data obtained from the experimental work. Best fitted curve was reported for each of the three different parameters.

6.3.1 Effect of Different Parameters on the Curing of PFA-Based Sealants

Here, a report is given on the optimized temperature, thickness, and catalyst concentration above which blistering during curing of PFA takes places. The result is expressed in terms of percentage difference in weight. If the blistering in the cured PFA takes place then there will be more increase in percentage difference in weight which is attributed to early setting-in of the exothermic reactions resulting in loss of less volatile compounds.

(a) Effect of thickness on the curing characteristics at constant temperature and constant amount of PTSA

Table 6.1 shows the effect of thickness on the curing characteristics of PFA sealants. The thickness was varied from 5 to 10 mm. The amount of catalyst concentration of 0.3 phr was kept constant in all the cases. As the thickness increased, the differences in the theoretical weight, i.e., weight of FA taken for polymerization and weight of PFA sealants after polymerization increased.

Figure 6.2 shows the photograph of PFA sealants. At lower thickness (5 mm), PFA sealants possess smooth surface but with the increase

Table 6.1 Effect of thickness on the curing characteristics at 50±2°C of furfuryl alcohol by incorporating constant amount of PTSA.

Thickness (mm)	Furfuryl alcohol (ml)	PTSA (phr)	Theoretical weight (gm) of FA before curing	Experimental weight (gm) of FA after curing	Difference in weight (%)
5	50	0.3	56.75	47.79	15.7
7.5	70		79.45	62.06	21.8
10	90		102.15	78.6	23.05

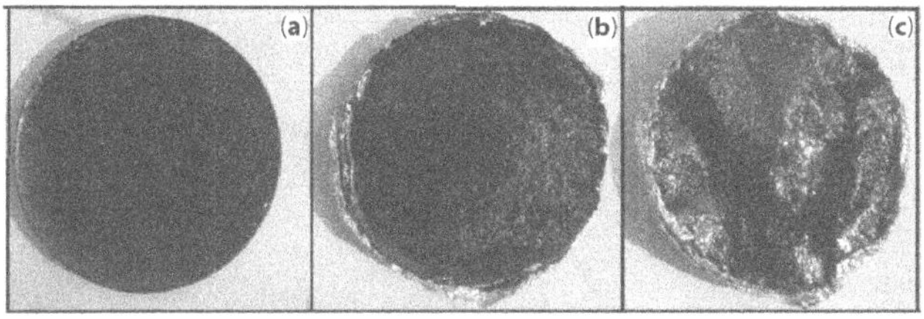

Figure 6.2 Photograph of PFA sealants at thickeness of 5 mm (a), 7.5 mm (b), and 10 mm (c) showing blistered and non blistered products.

in thickness from 5 to 10 mm, blistering in PFA sealants was clearly observed. This means that if the thickness of the PFA sealants increase to 10 mm, the curing reaction tends towards explosion and hence blistering takes place.

Using the curve fitting Toolbox in MATLAB software package, the values of the constants in the exponential model equation for curing of PFA with respect to thickness is given below.

$$f(x) = a^{*}\exp(b^{*}x) + c^{*}\exp(d^{*}x) \text{ ----------} \qquad (6.1)$$

$$a = 19.68; b = 0.01592$$

$$c = -1256; d = -1.082$$

The obtained values of a, b, c, and d were substituted into Equation (6.1). From the data, the percentage difference in weight wrt thickness was predicted. The predicted profile using the exponential is shown in Figure 6.3.

As can be seen in Figure 6.3, the exponential model equation gives relatively accurate percentage difference in weight at high thickness. The result in the following 2-by-2 matrix of correlation coefficient as obtained is given below.

1.0000	0.9042
0.9042	1.0000

From the above data, all correlation coefficients are close to 1. This means that there is a strong positive correlation between each pair of data in the columns as shown in the count matrix. From Figure 6.3, the profile validates the experimental data.

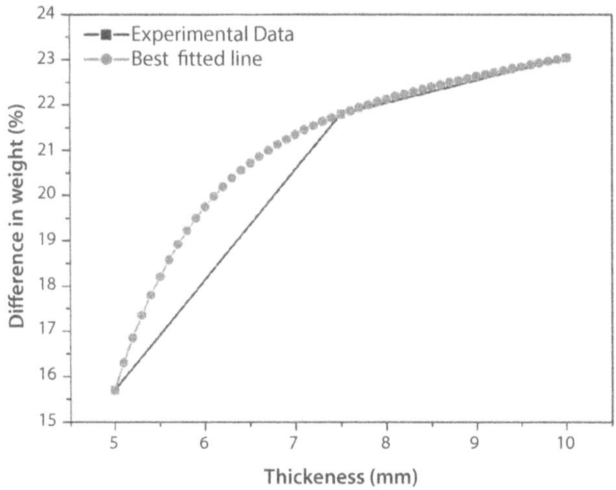

Figure 6.3 Comparison of predicted and experimental percentage difference in weight wrt thickness of the cured PFA-based sealants.

(b) Effect of catalyst content on the curing characteristics at constant temperature and constant thickness

Table 6.2 shows the effect of concentration of catalyst on the curing characteristics of PFA sealants. The concentration of catalyst was varied from 0.3 to 0.5 phr. The thickness of PFA sealants was kept constant in all the cases. As the concentration of the catalyst increases, there were no differences in the theoretical weight, i.e., weight of FA taken for polymerization and weight of PFA sealants after polymerization increased. The weight of cured PFA sealants, with different contents of PTSA, was in the range of 15±5 mg.

Table 6.2 Effect of concentration of catalyst on the curing characteristics at 50±2°C of furfuryl alcohol at constant thickness.

Furfuryl alcohol (ml)	PTSA (phr)	Thickness (mm)	Theoretical weight (gm) of FA before curing	Experimental weight (gm) of FA after curing	Difference in weight (%)
50	0.3	5	56.75	47.79	15.7
	0.4		56.75	47.14	15.7
	0.45		56.75	47.25	14.9
	0.5		56.75	47.51	14.5

Figure 6.4 shows the photograph of PFA sealants at different contents of PTSA. At lower concentration of catalyst, i.e., 0.3 phr, PFA sealants possess smooth surface but with the increase in concentration of catalyst from 0.4 to 0.5 phr, blistering in PFA sealants was clearly observed. This means that if the concentration of catalyst increases to 0.5 phr, the curing reaction tends towards explosion and hence blistering takes place [33]. Interestingly, in the experimental data there is percentage decrease in weight which is attributed to the fact that at low concentration of catalyst more time is needed for exothermic reaction to set in and hence more loss of volatiles.

Using the curve fitting Toolbox in MATLAB software package, the values of the constants in the Fourier model equation for curing of PFA with respect to amount of acid catalyst are given below.

$$f(x) = a0 + a1*\cos(x*w) + b1*\sin(x*w)\text{----------} \quad (6.2)$$

$$a0 = 10.79;\ a1 = 6.359$$

$$b1 = -6.78;\ w = 42.54$$

Equation (6.2) represents the best approximation to predict the percentage difference in weight as a function of content of acid catalyst. The predicted profile is shown in Figure 6.5.

The Fourier model equation in Figure 6.5 gives almost accurate percentage difference in weight at high content of acid catalyst. The result in the following 2-by-2 matrix of correlation coefficient as obtained is given below.

$$\begin{matrix} 1.0000 & -0.7121 \\ -0.7121 & 1.0000 \end{matrix}$$

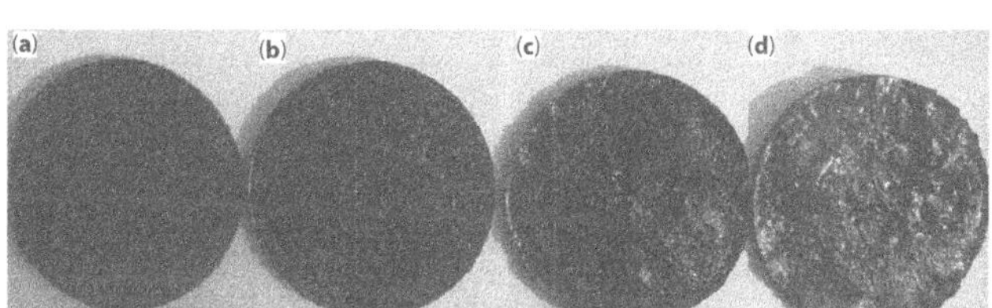

Figure 6.4 Photograph of PFA sealants at 0.3 phr (a), 0.4 phr (b), 0.45 phr (c), and 0.5 phr (d) of PTSA.

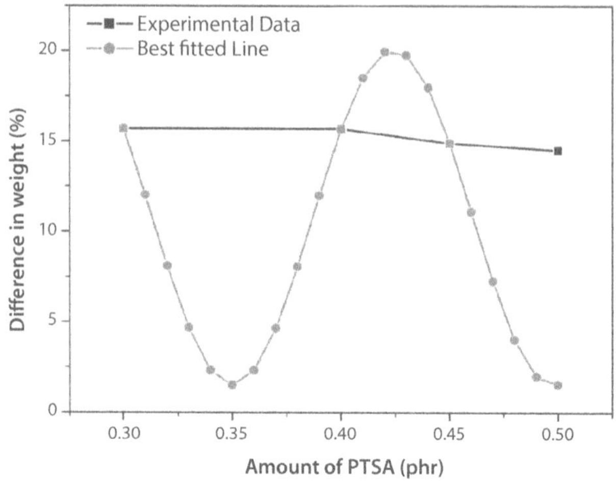

Figure 6.5 Comparison of predicted and experimental percentage difference in weight wrt amount acid catalyst.

From the above data, all correlation coefficients are not positive. This means that there is a negative correlation between each pair of data in the columns as shown in the count matrix. From Figure 6.5, the profile does not completely validate the experimental data except at high content of acid catalyst.

(c) Effect of temperature on the curing characteristics at constant thickness and constant amount of PTSA

Table 6.3 shows the effect of temperature on the curing characteristics of PFA sealants. The temperature was varied from 50°C to 60°C. The thickness of PFA sealants and concentration of catalyst was kept constant in all the cases. As the temperature increases, the weight of PFA sealant after polymerization decreased. This may be due to loss of PFA due to blistering.

Table 6.3 Effect of temperature on the curing characteristics of furfuryl alcohol at constant thickness and constant catalyst concentration.

Furfuryl alcohol (ml)	Temperature (°C)	Thickness (mm)	Theoretical weight (gm) of FA before curing	Experimental weight (gm) of FA after curing	Difference in weight (%)
50	50±2	5	56.75	47.79	15.7
	55±2		56.75	47.05	17.10
	60±2		56.75	46.40	18.14

Figure 6.6 shows the photograph of PFA sealants. At lower temperature, i.e., 50°C, PFA sealants possess smooth surface but with the increase in temperature to 60°C, blistering in PFA sealants was clearly observed. This means that if the temperature of curing increases to 60°C in the first step, the curing reaction tends towards explosion and hence blistering takes place [33].

Using the curve fitting Toolbox in MATLAB software package, the values of the constants in the linear model polynomial equation for curing of PFA with respect to temperature is given below.

$$f(x) = p1{*}x^3 + p2{*}x^2 + p3{*}x + p4 \text{----------} \quad (6.3)$$

$$p1 = -0.002905; p2 = 0.4793$$

$$p3 = -26.04; p4 = 482.8$$

The obtained values of p1, p2, p3, and p4 were substituted into Equation (6.3). From the data, the percentage difference in weight wrt temperature was predicted. The predicted profile is shown in Figure 6.7.

As can be seen in Figure 6.7, the linear model polynomial model equation gives relatively accurate percentage difference in weight at almost given range of temperatures. The result in the following 2-by-2 matrix of correlation coefficient as obtained is given below.

1.0000	0.9864
0.9864	1.0000

From the above data, all correlation coefficients are close to 1. This means that there is a strong positive correlation between each pair of data

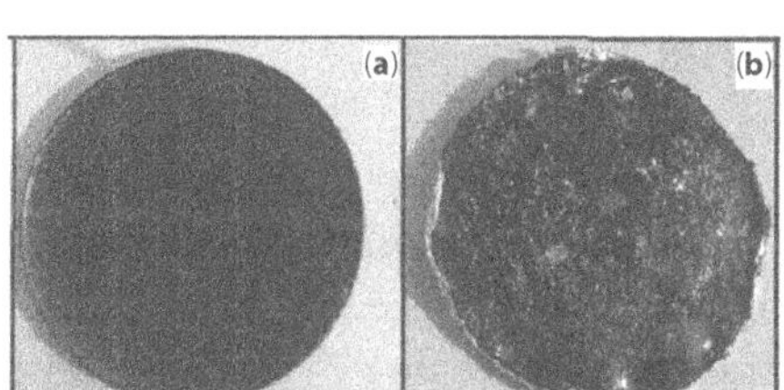

Figure 6.6 Photograph of PFA sealants at 50°C (a) and at 60°C (b).

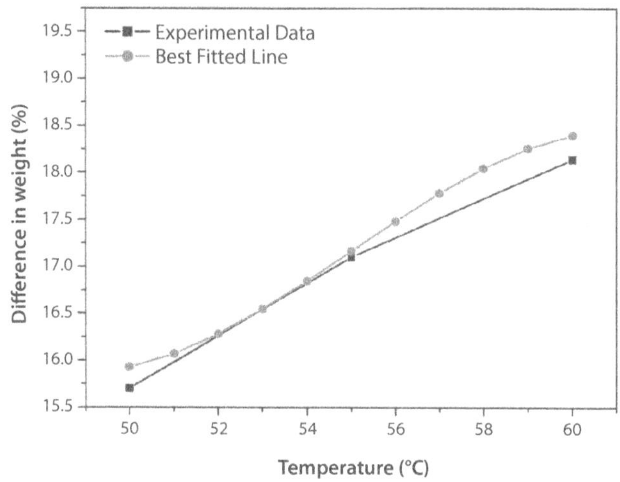

Figure 6.7 Comparison of predicted and experimental percentage difference in weight wrt temperature maintained during curing of PFA-based sealants.

in the columns as shown in the count matrix. From Figure 6.7, the profile validates the experimental data.

For this parameter, exponential fit, as given below, was also explored.

$$f(x) = a^{*}\exp(b^{*}x)\text{----------} \tag{6.4}$$

$$a = 8.865$$

$$b = 0.01171$$

The values of a and b were substituted into Equation (6.4) to predict the percentage difference in weight wrt temperature. However, it was not the best fitted curve.

6.4 Applications

Synthetic polymer-based adhesives in liquid form are widely used to fabricate particleboard and medium density fiber board (MDF). Synthetic polymer-based sealants are widely used in several sectors such as aviation, household, and automobile to provide structural integrity. Since synthetic polymer sealants are non-conductive in nature so in electronic industries, there are ample uses of these sealants. Synthetic polymer sealants provide effective bonding and are stable at high temperature. Polymer-based sealants are also used to seal concrete joints and metal joints. FA-based adhesives and sealants are cross-linked polymeric resins that are cured using heat and/or heat and pressure. FA-based adhesives can be available as one- or two-component systems. For one component systems, FA should be kept at low temperature to increase the shelf life. If FA adhesive is used as two component systems, then it must be mixed and applied within a set time frame, ranging from a few minutes to hours. FA behaves as a liquid, at some time in the course of bond formation, in order to flow over and wet the adherends. They must work with other components of the assembly to provide a durable product. After some time in presence of acid catalyst and at appropriate temperature, it hardens to carry sometimes continuous and variable load throughout the shelf-life. FA in polymerized form as PFA can fill gaps, cavities, and spaces.

In this book chapter, the model predictions were validated with the primary measurements of temperature profiles and cure. So, a validated process model can be used as an effective tool for process optimization. Appropriate parameter setting for curing reaction will eventually improve the biopolymer sealant quality, i.e., free from blistering. The application of PFA-based final products can be fruitful before blistering in the polymer sets in. After blistering, there may be chances of voids in the biopolymer sealants.

6.5 Conclusions

FA-based adhesive can be prepared in presence and absence of spent liquor and tannin. The resulting adhesive can be used to prepare particle board and to bond solid wood at 180°C. From FA one can easily prepare PFA-based sealants in presence of acid catalyst. But, the challenge is to prepare PFA-based sealants without blistering. PFA-based sealants without blistering can be fabricated provided one controls temperature of curing as well as thickness of PFA-based sealants. The optimum temperature for the first stage of curing was found to be 50±2°C and the optimum thickness was found to be 5 mm. Even at high concentration of acid catalyst (more than 0.3 phr), one can see blistering in PFA-based sealants.

Our studies on mathematical model validation of the curing process with respect to difference in percentage weight demonstrate the ability of models to follow the Exponential, Fourier, and Linear Polynomial equations for thickness, amount of catalyst and temperatures, respectively. Best fit curve was observed for Exponential and Linear Polynomial equations. The book chapter presented here provides insights into the curing process with respect to temperature, amount of the catalyst, as well as thickness of the end products.

Acknowledgement

The authors are grateful to the University Grant commission (UGC), New Delhi, India for granting us the Start-Up-Grant-Research-Fund (No. F 20-2(19)/2012 (BSR)). The part of this work has been delivered at the oral presentation in the conference (ICCMATS 2014).

References

1. Ramasamy, E., Karthikeyan, kumar, K.R., Veerappan, G., Glass-to-glass encapsulation with ultraviolet light curable epoxy edge sealing for stable perovskite solar cells. *Mater. Lett.*, 250, 51, 2019.
2. Wang, W., Kannan, P., Xue, J., Kannan, K., Synthetic phenolic antioxidants, including butylated hydroxytoluene (BHT), in resin-based dental sealants. *Environ. Res.*, 151, 339, 2016.
3. Fink, J., *Reactive Polymers: Fundamentals and Applications, Third edition*, pp. 225–253, Elsevier, New York, USA, 2018.
4. Jensen, M.K., Love, B.J., Grant, J.W., Cotton, J., Wilson, D.F., Comparison study of dicyandiamide-cured epoxy bonded steel joints and polyamidoamine-cured epoxy bonded steel joints. *Int. J. Adhes. Adhes.*, 20, 437, 2000.
5. Song, F., Tang, D.-L., Wang, X.-L., Wang, Y.-Z., Biodegradable soy protein isolate-based materials: A review. *Biomacromolecules*, 12, 3369, 2011.
6. Kumar, R., Liu, D., Zhang, L., Advances in proteinous materials. *J. Biobased Mater.*, 2, 1, 2008.
7. Majid, J., Elmira, A.T., Muhammad, I., Muriel, J., Stéphane, D., Poly-lactic acid: Production, applications, nanocomposites, and release studies. *Compr. Rev. Food Sci. F*, 9, 552, 2010.
8. Pranger, L. and Tannenbaum, R., Biobased nanocomposites prepared by in situ polymerization of furfuryl alcohol with cellulose whiskers or montmorillonite clay. *Macromolecules*, 41, 8682, 2008.
9. Choura, M., Belgacem, N.M., Gandini, A., Acid-catalyzed polycondensation of furfuryl alcohol: Mechanisms of chromophore formation and cross-linking. *Macromolecules*, 29, 3839, 1996.
10. Trindade, W.G., Hoareau, W., Razera, I.A.T., Ruggiero, R., Frollini, E., Castellan, A., Phenolic thermoset matrix reinforced with sugar cane baggase fibers; Attempt to develop a new fiber surface chemical modification involving formation of quinones flowed by reaction with furfuryl alcohol. *Macromol. Mater. Eng.*, 289, 728, 2004.

11. Babu, R.P., O'Connor, K., Seeram, R., Current progress on bio-based polymers and their future trends. *Prog. Biomater.*, 2, 8, 2013.
12. Choudhary, V., Mishra, S., Varma, I.K., Mattiason, Bo., Adhesive and plastics based on soybean products. *Ind. Crops Prod.*, 16, 155, 2002.
13. Ghaffar, S.H. and Fan, M., Lignin in straw and its applications as an adhesive. *Int. J. Adhes. Adhes.*, 48, 92, 2014.
14. Elvin, C.M., Alan, T.V., Brownlee, G., Sando, L., Huson, M.G., Liyou, N.E., Stockwell, P.R., Lyons, R.E., Kim, M., Edwards, G.A., Johnson, G., McFarland, G.A., Ramshaw, J.A.M., Werkmeister, J.A., A highly elastic tissue sealant based on photopolymerised gelatine. *Biomaterials*, 31, 8323, 2010.
15. Gandini, A., Polymers from renewable resources: A challenge for the future of macromolecular materials. *Macromolecules*, 41, 9491, 2008.
16. Gandini, A., Furans as offspring of sugars and polysaccharides and progenitors of a family of remarkable polymers: A review of recent progress. *Polym. Chem.*, 1, 245, 2010.
17. Deka, H., Misra, M., Mohanty, A., Renewable resource based "all green composites" from kenaf biofiber and poly (furfuryl alcohol) bioresin. *Ind. Crops Prod.*, 41, 94, 2013.
18. Kumar, R. and Anandjiwala, R.D., Polyfurfuryl alcohol products, South Africa Patent Application ZA2011/01976, filed by CSIR, South Africa, 2011.
19. Kumar, R. and Anandjiwala, R.D., Process for preparing polyfurfuryl alcohol products, WIPO, WO 2012/123902 A1, filed by CSIR, South Africa, 2012.
20. Kumar, R., Aluminium/iron reinforced polyfurfuryl alcohol resin as advanced biocomposites. *AIMS Mater. Sci.*, 3, 908, 2016.
21. Jang, B.Z. and Shih, W.K., Techniques for cure monitoring of thermoset resins and composites—A review. *Mater. Manuf. Process.*, 5, 301–331, 1990.
22. Lionetto, F. and Maffezzoli, A., Monitoring the cure state of thermosetting resins by ultrasound. *Materials*, 6, 3783, 2013.
23. Hornik, K., Stinchcombe, M., White, H., Multi-layer feed forward networks are universal approximators. *Neural Netw.*, 2, 359, 1989.
24. Ito, Y., Approximation capability of layered neural network with sigmoid units on two layers. *Neural Comput.*, 6, 1233, 1994.
25. Liang, G. and Chandrashekhara, K., Cure kinetics and rheology characterization of soy-based epoxy resin system. *J. Appl. Polym. Sci.*, 102, 3168, 2006.
26. Srinivasagupta, D., Potaraju, S., Kardos, J.L., Joseph, B., Steady state and dynamic analysis of a bench-scale injected pultrusion process. *Compos. Part A: Appl. Sci. Manuf.*, 34, 835, 2003.
27. Win, D.T., Furfural—Gold from garbage. *AU J. Technol.*, 8, 185, 2005.
28. Kumar, R., Patents related to catalysts used for curing thermoset resins. *Recent Pat. Catal.*, 1, 35, 2012.
29. Kuster, B.F.M., 5-Hydroxymethylfurfural (HMF). A review focussing on its manufacture. *Starch Staerke*, 42, 314, 1990.

30. Salem, M.Z.M. and Bohm, M., Understanding of formaldehyde emissions from solid wood: an overview. *Bioresources*, 8, 4775, 2013.
31. Luckeneder, P., Gavino, J., Kuchernig, R., Petutschnigg, A., Tondi, G., Sustainable phenolic fractions as basis for furfuryl alcohol-based co-polymers and their use as wood adhesives. *Polymers*, 8, 396, 2016.
32. Abdullah, U.H.B. and Pizzi, A., Tannin-furfuryl alcohol wood panel adhesives without formaldehyde. *Eur. J. Wood Prod.*, 71, 131, 2013.
33. Yun, S., Wei, L., Minghui, Z., Keyi, T., Hydrogenation of furfuryl alcohol to tetrahydrofurfuryl alcohol on NiB/SiO_2 amorphous alloy catalyst. *Front. Chem. Eng. China*, 1, 151, 2007.

7
Bioadhesives

M. Ramesh[1]* and L. Rajesh Kumar[2]

[1]Department of Mechanical Engineering, KIT-Kalaignarkarunanidhi Institute of Technology, Coimbatore, India
[2]Department of Mechanical Engineering, KPR Institute of Engineering and Technology, Coimbatore, India

Abstract
Bioadhesion is a process occurring in natural and synthetic materials with the biological surfaces. In order to develop technologies like biosensors and some other novel materials and process, new age researchers focus keenly to develop a familiarity over the governing mechanism of bioadhesives. Mechanism of bioadhesion is used by various organisms for applications like defense, construction, etc. Meanwhile, medical operations increase their focus on bioadhesion mechanisms like sealants, adhesives, and haemostatic proxies. During the past decades protein bioadhesives are considered to be highly suitable biocompatible alternatives for synthetic bioadhesives. In technical applications, protein-based bioadhesives are widely used by combining biochemical and mechanical principles. For such applications knowledge on design, composition and surface interaction of adhesives are required. Even in after-surgical injuries and normal tissue rehabilitation functions bio-based adhesives and glues were used traditionally. It is believed normally that the use of bioadhesives in current trend not only increases the environment sustainability but also the reliability of adhesion when compared with the synthetic adhesion. This chapter deals with the overview of bioadhesives, their history, and classification. Light has also been thrown upon the mechanism of bioadhesion along with the testing methods and application of bioadhesives.

Keywords: Bioadhesives, bioadhesion, biochemical, synthetic adhesion, biocompatibility

**Corresponding author*: mramesh97@gmail.com

Inamuddin, Rajender Boddula, Mohd Imran Ahamed and Abdullah M. Asiri (eds.) Green Adhesives: Preparation, Properties and Applications (145–164) © 2020 Scrivener Publishing LLC

7.1 Introduction

Any substance that can undergo polymerization can be termed as an adhesive or glue. This polymerization may function in cross-linking two surfaces together or seal the blocks at all possible locations where they are employed [1]. Wood, glass, and papers can readily be termed as adherents since they are directly associated with the adhesives. If any of the above adherents are in combination with the some natural materials, then they may be termed as bioadhesives [2]. Bioadhesion can be defined as the phenomenon of holding two materials together for a prolonged period of time by means of interface tailoring. Biocompatibility, biodegradability, and large molecular weight are the humongous merits of the bioadhesives. Apart from the specific characteristics of the bio-organisms, bioadhesion is mostly similar phenomenon like traditional adhesion [3]. Bioadhesion covers the adhesion properties that bonds two surfaces together and many applications like dental, surgery, and drug delivery systems utilize these principles.

Considering many facets of bioadhesion, ecological point of view is to focus mainly in order to achieve resource efficiency, mainly for which the bioadhesives are used these days. As bioadhesives contain constituents either form natural or synthetic origin, they are widely used as adherent materials in variety of applications that involves surgery time reduction, strengthening of seals, less requirements for removal materials, application easiness, sealing air leaks with enhanced quality, and so on. A bioadhesive should be characterized by non-immunogenity, bio-degradability, non-toxicity, better strength with effectiveness in physiological conditions [4].

Application of bioadhesives earned much popularity in clinical operations like sealants, haemostats, and glues which gained success in these fields. Its objective is to hold the tissues during healing of injuries and avoid the inclusion of foreign bodies at the damaged surfaces [5]. Meanwhile, bioadhesives should act at specific locations and progress with healing process with utmost safety. Bioadhesion in wet surfaces, which is considered to be challenging task in adverse environment, was easily solved by marine organisms like mussels, fungi, and other bacterium [6]. Marine mussels are the natural organisms possessing inherent ability to adhere to other surfaces by using their natural adhesive proteins. The bonding plaques, which are called Mussel Foot Prints (MFP), created by these mussels at the interface are tough, durable, and strong. The primary ingredient of MFP is 3, 4 Dihydroxyphenylalanine (Dopa) which is the derivative of tyrocine. MFP obtains the property of binding and solidifying with substrates through catechol side chain of Dopa by means of cross-linking and chemical reactions. Novel adhesive materials were currently developed based on the MFP to cater the needs of wet adhesions by reaping the benefits of using adhesive proteins [7].

Aristotle rightly pointed out that "If there is a better solution, then it would be available with the nature itself". Likewise, naturally flora and fauna possess the ability of adhesion to substrates and tissues which serves as the motivation behind development of sustainable bioadhesives [8]. Such bioadhesives are utilized in numerous applications and to develop technologies to obtain high performance that are on par with the nature.

Blood coagulation process can also be replicated by the bioadhesives using proteinaceous and polysaccharide materials. Fibrin sealants, gelatin, and collagen are the protein-based adhesives, while alginate, chondroitin, and chitosan sealants fall under polysaccharide types [9]. Bioadhesion concept can also be used for drug delivery system (DDS) through drug carriers to particular locations. Bioadhesion can take place between two surfaces upon which the substrate may be a mucous or epithelial tissue. If the bioadhesion is with the former, then it is termed as mucoadhesion. Mucoadhesion linings may include locations like oesophageal, vaginal, oral, nasal, rectal, or ocular cavities [10].

Inspite of the numerous advantages of the bioadhesion, it may also cause negative effects such as biofouling. If the formation of biofilm happens on solid surfaces like the body of animals or human, due to clustering of micro-organisms or other biomaterials, then it is called disease. Diseases

Figure 7.1 Biofilm formation in polyurethane stent [13].

like cystic fibrosis and endocarditis is believed to be caused by bacterial biofilm which usually have greater resistance towards antibiotics [11]. It is believed that the dental biofilm formed over teeth and gums may have over 500 type of bacterium [12]. A biofilm formation over the medicated polyurethane stent placed in the human body is shown in Figure 7.1 [13]. Current chapter deals with the overview of concepts like bioadhesives and bioadhesion, history of the bioadhesives and their classification, mechanism of bioadhesion, and the testing methods for the strength of bioadhesive formation between various substrates.

7.2 History of Bioadhesives

Creating bonding between numerous layers or tissues is the traditional technique of adhesion or gluing mechanism. If one of the adhering elements is natural, then the binding is called bioadhesion. Bioadhesives may contain lipids, carbohydrates, protein, or any one among them in varying proportion. Neanderthals recorded the most primitive application of bioadhesives before 2000 centuries. Archaeologists also cited this use of bioadhesive during the middle and towards the end of Palaeolithic era. Considering the most practical aspect, ochre was the first natural bioadhesive employed during the above mentioned period. Post this usage, at war times, that obtained from the birch bark was used to join the stones to the sword blades [14]. Over 1,000 years, Greeks, Romans, and the Egyptians used various adhesives like semiliquid balsams, wet lime, tree resins, and collagen to join materials like ceramics, marquetry, and wooden plies. Hence, it could be directly quoted that the technology of using adhesives in marquetry and joining wooden plies were invented by Greeks and Romans. Apart from the above natural plant elements, fishes, and animal-based glues also originated from the mother nature. But the advent of plastic adhesives was very high during twentieth century particularly at the period of industrial revolution [15]. Figure 7.2 depicts the usage of bioadhesives for various purposes like protection, movement, defense, and attachment.

Fibrin-based bioadhesives were the topics of study since 1940 in terms of extraction of fibrinogen from plasma of human blood. Its utilization mostly was adhesive for peripheral nerves. In order to fully enjoy the benefits of fibrin glue, commercialization and approval of fibrin-based sealants were done by Europe in 1972 and Food and Drug Administration of US in 1990 respectively. During recent times, fibrin sealants are at an advance stage of study and at the fullest utilization as bioadhesives [16].

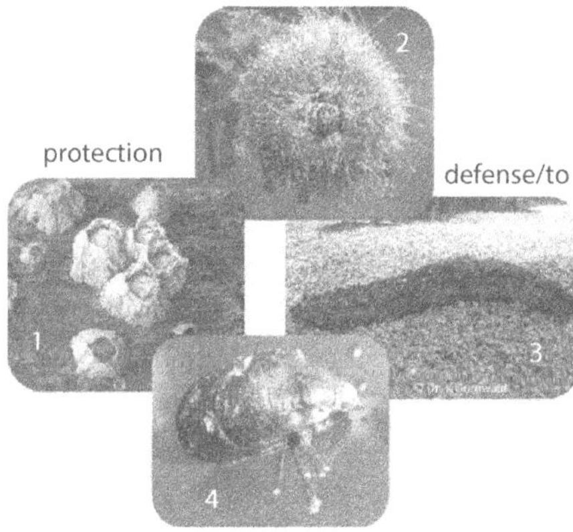

Figure 7.2 Bioadhesives used for different purposes [19].

When compared with other joining techniques, bonding by using synthetic adhesives is very cheaper and economic, due to which the production of those adhesives is very high in current days. Usage of adhesives is very common in fields like artistic, engineering, building, and construction applications like wall covering and flooring. One promising area where adhesives are used is the timber building where many layers of timber plies can be bound by adhesives to produce customized products. This is very predominantly happening in USA and some other countries. Additionally, the timber construction market is equally rowing with the petrochemical industries, but the latter is depleting gradually due to the increased public awareness [17]. Interests by researchers grow rapidly in developing bioadhesives for research and industry. This is evident from the following data: A total of 30 research articles were published in the year 1990 while it has increased to a huge count of 1,200 numbers in the year 2012. This can be directly correlated with the increasing awareness for building health-friendly bioadhesives [18]. Recently, biomimetic and bioinspired materials are being developed and are now reaching a new height of advancements.

7.3 Classification of Bioadhesives

Bioadhesives are classified as external and internal bioadhesives based on their function and compatibility. For a bioadhesive to be tagged as internal, it has to satisfy two basic criteria: primarily, it should be soluble in aqueous solution without addition of any organic solvent to the primary constituents. Secondly, the primary adhesive constituents must be able to undergo cross linkage. These conditions are formulated due to the fact that aqueous solutions contain minimum toxic content and internal bioadhesives will frequently be in contact with internal organs and fluids [20]. Internal bioadhesives are believed to be highly compatible with intracorporal environment when compared acetic acids which are non-aqueous nature. It is also a fact that bioadhesives can establish their adhesive function only when it is cross-linkable in wet conditions as the internal organs would have fluid circulation with rich supply of blood. Long-term toxicities and side effects may occur in the recipient's body, if the bioadhesives fed inside are incapable of dissolving and degrading in bodily fluids which are sent out by excretion [21]. When any of the above said conditions could not be satisfied, then they can be classified and used as external bioadhesives. If any of the bioadhesive is proven, by any means, that it is not compatible in intra-corporal conditions, then it fails to be eligible for usage as internal bioadhesives, but can be used as an external bioadhesives in direct contact with the external atmosphere [22].

Most widely used type of external bioadhesive is the cyanoacrylate-based tissue adhesive which goes a long way back in history. Some of those adhesives like Trufill n-BCA and Dermabond are utilized in variety of applications like plastic surgeries, wound dressings and treatment, and skin transplantations. All these adhesives are approved by Food and Drug Association (FDA), USA. Cyanoacrylates are characterized by advantages like short time for adhesion and better bonding strength. Nevertheless, cyanoacrylates end up in poisonous elements after degradation such as formaldehyde and their respective alkyl compounds due to which their usage is limited to the tropical zones. Apart from the regular complications like necrosis, thrombo-embolic, and septic complications, these toxic elements may also act as carcinogenic agents. If the brittleness caused by the shorter length of alkyl groups after the adhesion process and if the decrease in adhesive strength due to longer chains of alkyl group is addressed properly, cyanoacrylates may find expansion in terms of applications [23].

7.4 Mechanism of Bioadhesion

Phenomenon of adhesion through bioadhesives differs from the traditional adhesives as per the requirements of the substrate properties and interfacial characteristics. When a contact is established strongly between a liquid or solid phases with respect to molecular adhesion, then it can be termed as adhesion [24]. This can be also termed as chemical mechanism when the adhesion follows electrostatic theory and adsorption theory or as physical mechanism when it follows diffusion and interpenetration [25]. When the adhesive wets the entire surface and if a physio-chemical intermolecular force develops between the substrates and adhesive, it may also be termed as adhesion. As far as both the surfaces joined by adhesive are concerned, the surface energy is vital as the adhesion mechanism follows wetting theory which utilizes the interfacial tensile forces to forecast the bioadhesive's ability [26].

Bioadhesion is basically divided into two types: specific and nonspecific adhesions. Nonspecific adhesion involves binding of the adhesives to the surface irrespective of the surface properties. For example, polymers that adhere to most of the surfaces exhibit these properties. Specific adhesion involves binding of the adhesive by recognition of the specific target groups on the cell surface. Another example is tomato lectin that binds specifically to the surface via linkage with N-acetyl glucosamine [27].

7.4.1 Mechanical Interlocking

Figure 7.3a shows the process of infiltration of bioadhesives into the surface anomalies and pores of the substrate material. If the adhesive fills and intertwines mechanically with the surfaces at micro level then binding

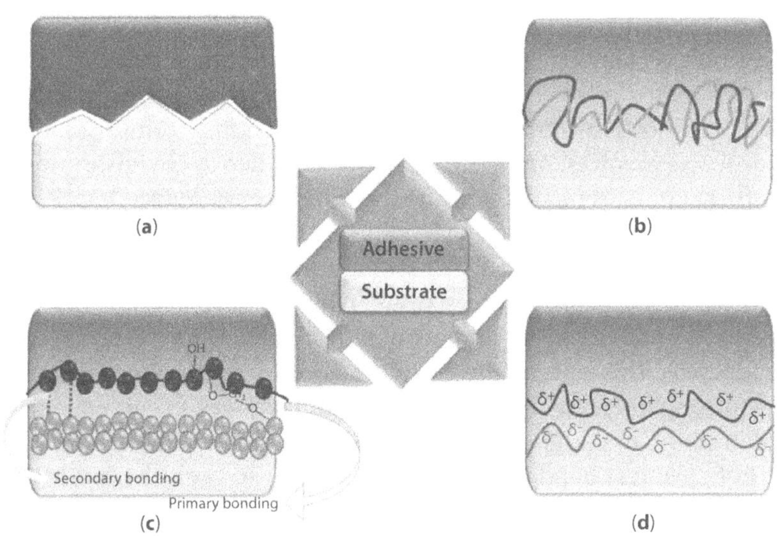

Figure 7.3 Mechanisms of bioadhesion [29].

of the surfaces takes place. In order to ensure better surface bonding, pre-treatment may be necessary in some cases. But this mechanism has a lower probability of working out when compared with smooth surface interlocking which has also been reported in earlier studies. Naturally, smooth surface interlocking happen between the tooth surface and amalgams by means of mechanical intertwining and even the tooth surface is considered to be pre-treated [28].

7.4.2 Chain Entanglement

Any adhesion happening between two different polymers, either similar or dissimilar in nature, with respect to molecular mobility is termed as chain entanglement. Due to impossible molecular mobility in large-scale, it is clear that the phenomenon has no possibility to take place in a highly amorphous or highly crystalline polymers or in cross-linked polymers particularly beneath the temperature of glass transition. A layer of entangled polymer chain arises due to rambling of macromolecular polymers conjointly within the existing polymer molecules and its contact surface with the substrate. This is depicted in Figure 7.3b [30]. For instance, the entanglement and interweaving of mucus glycoproteins and mucoadhesive layers in a mucoadhesive DDS is the primary causal factor of adhesion [31].

7.4.3 Intermolecular Bonding

As a result of binding of bioadhesives and substrate elements intermolecular bonding arises. This is the predominant mechanism of adhesion between various materials. Intermolecular bonding comprises of two major forces involves: one is the primary force due to chemisorption and the other is the secondary forces. This is illustrated in Figure 7.3c. Primary forces are liable to produce tough adhesion and high energy due to some major bonds including metallic, ionic, and covalent ones. But, the secondary forces, unlike primary forces, result in lower energy levels which can be increased by increasing the number of interfacial locations of substrate and the adhesive. This in turn increases the bond strength also. These forces contain some forces like dipole interaction, hydrogen bond, London dispersion, and Van der Waals forces. With respect to bioadhesion, establishment of chemical bond has to be predominant to bind the molecules of tissue surface with the adhesives and this bonding may arise from primary or secondary forces or due to both [32].

7.4.4 Electrostatic Bonding

When the surface of the substrate molecules and adhesives are brought closer to each other, then exchange of electrons between them generates electrostatic forces. This results in formation of twofold layer of electrons at the interface due to stability achieved in Fermi levels. This phenomenon is shown in Figure 7.3d [33]. Formation of electrostatic charge is the basic concept of bioadhesion. However, in insulating materials substrates, the time for charge build-up is high since the available number of electrons for exchange is a major constraint. For such materials the charge concentration in the interface is very minimum as observed in between the adhesion of mucus glycoprotein and bioadhesive material used [34].

7.5 Testing of Bioadhesives

Usually two types of tests were carried out to evaluate the properties and characteristics of bioadhesives, namely, *in vitro* and *ex vitro* methods. The detailed description of these methods is given below.

7.5.1 *In Vitro* Methods

7.5.1.1 Shear Stress Measurements

Conventional mechanical characterization methods including shear test, peel test, and crack growth assessment can be used to assess the strength of bioadhesion. Shear stress measurements are normally used to measure the sliding force in the mucus layer which slides, in a parallel direction to contact plane, on the mucoadhesive. Flow channel method is usually used to measure the shear strength of the mucoadhesive. It measures the shear strength by measuring the force required to isolate an adhesive particle from the mucin gel surface by means of forced humid air via flow cell [35].

In dental-based bioadhesion mechanisms, bending tests were conducted to evaluate the growth of crack originating from the dental implant. This is due to the shrinkage of the composite materials used in the implant as a result of polymerization. Griffith's energy balance model is used for the characterization and interpretation of the bending test results. This balancing model tries to establish a balance between two energy levels such as the teeth elastic energy (usually the average elastic energy of tooth and the dental implant material) and the crack surface energy. Strain energy release rate or the stress intensity can be determined from the experimental crack growth evaluation and the fracture energy can be calculated from the Poisson's ration and modulus of the implant material [36].

7.5.1.2 *Peel Strength Evaluation*

In order to assess the quality of the dental implants-surface, after making them to undergo tensile tests, these are usually characterized by fractographic techniques like transmission electron microscopy (TEM) or scanning electron microscopy (SEM). Various tests that are conducted on the interface of adhesion and the substrates along with the American Standards for Testing and Materials (ASTM) are listed in Table 7.1 [37]. Normally, pressure-sensitive adhesives (PSA) is formed as a composite material by adding it with montmorillonite, an organo-clay based element, for obtaining better shear strength, peel strength, and adhesion failure temperature [38].

7.5.1.3 *Flow Through Experiment and Plate Method*

Mucoadhesion of DDS is usually measured by the flow through channel method which is the macroscale measure of flow rate which can result in the depletion of bioadhesive coated over the substrate sphere. An analytical centrifuge measures the fluctuation in sedimentation coefficient that arises due to the molecular weight change which is continuously observed by a biophysical assessment method. Yet another macroscale bioadhesion assessment method is Wilhelm plate method which utilizes natural or synthetic mucus instead of a conventional water medium for surface tension evaluation. This is done over a plate which is coated with any polymer material and the results would be the interfacial properties and the bioadhesion property changes with respect to time [39]. Figure 7.4 schematically represents various macro level testing methods of bioadhesives.

Table 7.1 Tests and their corresponding ASTM standards.

S. no.	Test description	ASTM standards
1	Tensile test	ASTM D368
2	Shear strength	ASTM D3654
3	Peel test	ASTM D903
4	Shear adhesion failure temperature (SAFT)	ASTM D4498

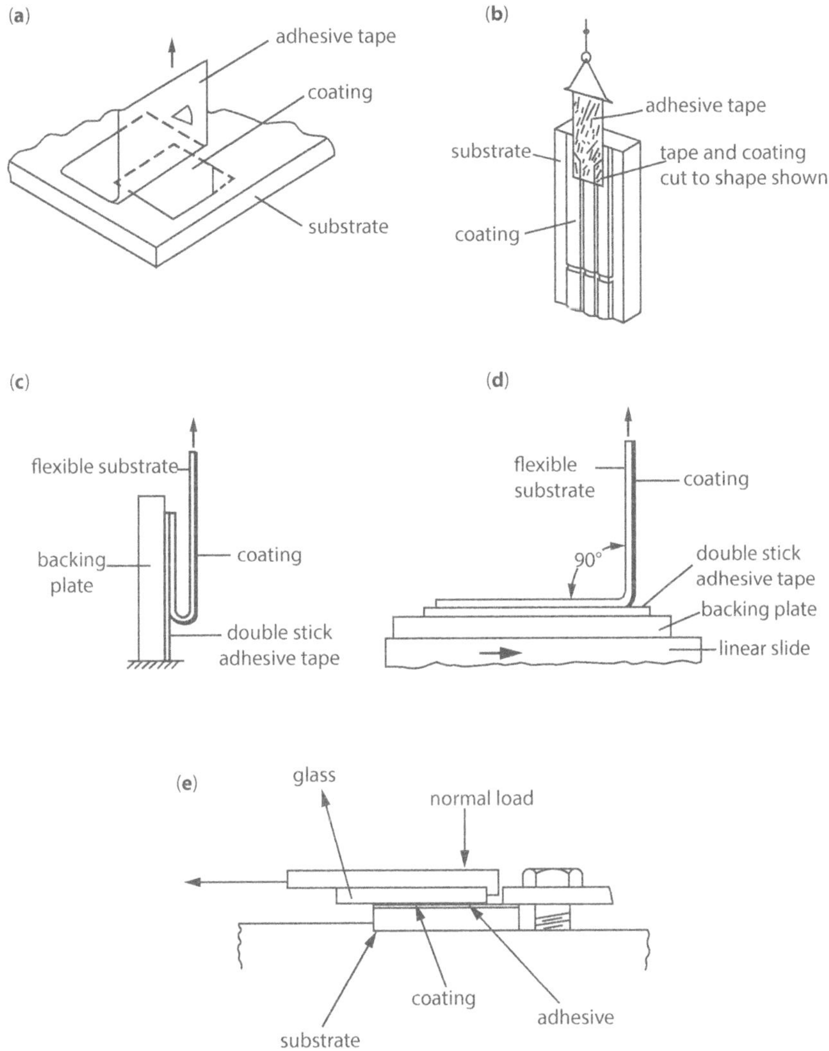

Figure 7.4 Macrolevel testing of bioadhesives [39].

7.5.2 *Ex Vitro* Methods

7.5.2.1 Adhesion Weight Method

A specific test method which determines the weight of adherent particles that arose due to the ion exchange in the interior mucous layers of guinea pig intestine was developed. This method was capable of determining the particle size effect and adhesion charge after 5 min of time with the pig's intestine. Results, with larger sieve size ranging between 63 μm and 178 μm, suggested that the weight of the intestine increased due to bioadhesion. But, the major limitation of this method is that it posed a reduced reproducibility of the data when a larger change in the biological tissue arose due to regeneration or degeneration of the intestinal tissues [40].

7.5.2.2 Fluorescent Probe Methods

Relationship between the polymer molecules and epithelial cell membranes were analyzed through fluorescent probe methods. Prime focus of the study was to develop an improved formulation of orally used bioadhesive polymer by knowing the structural requirements. Proteins and the lipid bilayer membranes were tagged with the bioadhesive under evaluation and the variations in fluorescent spectrum are noted. Pyrene bands of these materials exhibited two different bands, namely, excimer and monomer bands while the ratio of these bands was governed by the environmental viscosity. Hence, it is sufficient if the viscosity changes were noted by assessing the bands ratio. From this result, it was assumed that the viscosity change can be directly related with the strength of adhesion. Fluorescence depolarization was a quantitative method that was used to observe the bond between polymer and protein membrane wherein the interactions of soluble polymers can be compared with that of peel of the cell [39, 41].

7.5.2.3 Falling Liquid Film Method

Figure 7.5 shows the falling liquid film method which used micro sized particles and excised segments. For this method 1% of polymer particles were added into 5×10^8 particles/ml of clean latex particles, mixing them up in water by stirring for at least 2 h. In order to enhance the effectiveness of coating by varying the ionic strength using buffer solutions, the prepared mixture was transferred to the target buffer solution to end up with 5×10^6 particles/ml. This suspended mixture was sonicated for 30 s and used after a delay of 15 min.

Then, the small intestine of Sprague-Dawley rats was taken, cut into pieces along their length using surgical scissors, cleansed using saline water and spread over 1" internal diametric Tygon tubes. These tubes were placed over plastic foam bed and their angle of inclination was set to 78°. Loose mucus was removed from the intestine by using a perfusion pump after the cut intestines were spread over the Tygon plates. Then, after 2 min delay of coating the intestines with the prepared buffer solution, sample

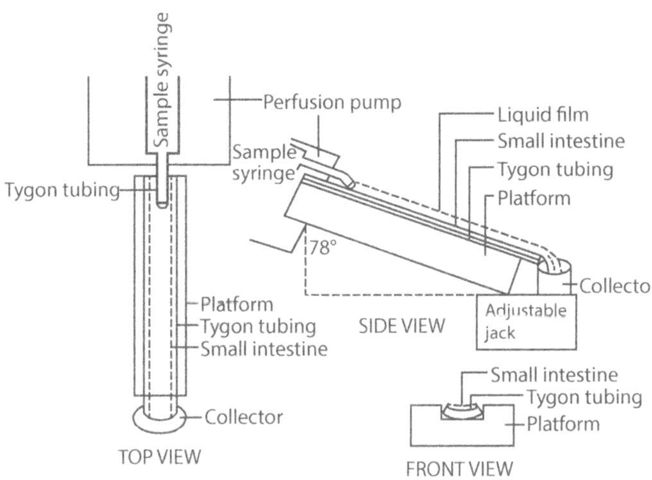

Figure 7.5 Schematic of falling liquid film system [42].

solutions were collected from the field to commence the counting of particles. It was noted that an extraneous substance was present on the surface of the intestine which may be mucus. For measuring the bioadhesion phenomenon over the particle coated intestinal surface, 0.5 ml of samples were recovered and using an Coulter particle counter the number of particles was assessed [42].

7.6 Application of Bioadhesives

7.6.1 Bioadhesives as Drug Delivery Systems

For targeted delivery system, bioadhesive polymers that interact with specific targets are the key factor. If they lack specificity, then the targeted DDS may not reach the targets, say gastro intestinal (GI) tract. This lacuna may end the oral drug in the first mucosal layer which may result in local tissue damage due to wrong drug administration [43]. Another major problem is that, if specificity is absent for the delivery material, then the drug would be coated with the loose mucus present around and passes through the actual mucosal membrane without getting absorbed. Molecules that possess specific adhesion are invasions, lectins, and fibrins. Various cells are characterized by matching amino acids in mucosal surfaces and may enhance the bond with surface glycoproteins when connected with micro particles [44]. Thus, by using those matching amino acids in DDS, the changes in surface glycoproteins at the time of diseases can be embattled.

For chronic diseases like tumors, these high specific DDS may be attached with antibodies or designed with location specific mucoadhesives.

7.6.2 Bioadhesives as Fibrin Sealants

Any design of fibrin sealant formulations may contain only two constituents, namely, thrombin and fibrinogen. When these sealants are exposed to the bleeding locations, they render a better interaction and establish an unstable clot of fibrin that possesses better cohesive strength due to hydrogen bonding. This unstable layer may further cross-link with the target tissues and enhances the adhesive strength. Some of the common areas of application of fibrin sealants are in skin grafting, haemostatics, and colonic sealing which are labeled by FDA. Clinical operation like cardiovascular surgeries, hernia repair, and seroma prevention are using fibrin sealants and are under experimental analysis. Biotin and avidin are vitamins and proteins, respectively, and the systems which have rich biotin-avidin interaction are predominantly used due to advantages like specific binding. Biotin is available in mammary gland, kidney, and liver tissues of all mammals and avidin is available in egg whites. Application of biotin-avidin systems includes B, T-cell antigen detection, membrane ultra-structural staining, quantitative enzyme-immunoassay, and gene mapping and enrichment. Bioadhesives derived from the biotin-avidin interactions are majorly used as surface coating to enhance the specificity of cell adhesion and substrates [45].

Fibrin glue was experimented as a scaffold for myocardial infarction (MI) by employing it as a support to the vessel wall of a rat. Results indicated that wall contraction could be avoided by using fibrin glue at the time of MI. Another experiment proved that the usage of fibrin glue minimized the infraction by enhancing the flow of blood to myocardium thereby increasing the neovascularization of the rat. It could also be used to maintain proper cardiac functioning after MI by using adipose-derived stem cells. Advantages of the combined use of fibrin glue and stem cells are enhanced heart functioning, increased arteriole density, enlargement of the graft size, and better cell retention [1].

7.6.3 Bioadhesives as Protein-Based Adhesives

Proteins comprise of amino acids which diverge in their filtrates. Depending on their composition, they can be tailored for different purposes ranging from movement (actin), catalysis (enzymes), fixation of nitrogen (nodule bacteria), structuring (collagens), protection (immunoglobulins), transport (ionic channels, hemoglobins), and communication (hormones) [46]. Gluten from wheat is a combination of gliadin (56%) and glutenin (44%) is a storage elastic protein with excellent adhesive strength and characteristics. Seitan consists of gluten as second major constituent next to water. It is mostly used in fields like additive to mortar and concrete, paper industry, and in adhesives. Another protein termed as casein is also widely used in painting the caves, painting sausage bottles, label printing, kindergarten paints, interior paints, and also as an adhesive. It is obtained by minimizing the pH value of the milk and is a thermoset. It can also be called as galalith when treated with formaldehyde. Casein paints can be converted to be resistant against water by incorporating chalk milk to it. Due to the high demand for food products, casein is least utilized nowadays [7].

7.6.4 Bioadhesives in Tissue Engineering

Application of bioadhesives currently in tissue regeneration and engineering has gained prominence, apart from its usage in wound healing and haemostatics. When the biomaterials are applied for tissue regeneration, then the interfacial gaps between the tissues cannot be filled by the bioadhesives properly, which isolates the subsequent tissues from binding with the bioadhesives. This is a major drawback noted in employing the tissue rebuilding. Yet this could be eradicated by the proper suturing mechanism and effective usage of bioadhesives. Need of bioadhesives may vary from tissue to tissue since the properties and functional requirement of the tissues vary. Hence, tailored tissue adhesives that focus on specific tissues were developed and employed to avoid the failure mishaps. In a recent research an attempt has been made to examine the possibility of employing the adhesive in cartilage tissue rebuilding for integration of the graft. This was made possible by crosslinking the chondroitin sulfate (CS), a cartilage polysaccharide, with aldehyde groups and methyl acrylates [47].

CS can be also used in cartilage tissue treatment as they are injected into the nearby tissues and are applied as biodegradable scaffold adhesive. Methylacrylate group of CS establishes bond with PEG diacrylate by means of photo linking and the aldehyde group establishes a chemical bond. Hence, cartilage rehabilitation could happen by establishing a bond between cartuilage tissue and the bioadhesive by means of graft integration. This is depicted in Figure 7.6. From the studies, it could be concluded that the cartilage rehabilitation takes place effectively by using bioadhesives coupled with hydrogels [48]. Fibrin glue is a widely tested and used bioadhesive in tissue engineering. Such use of fibrin glue may promote reduced mechanical inflammation, improved regeneration of tissues,

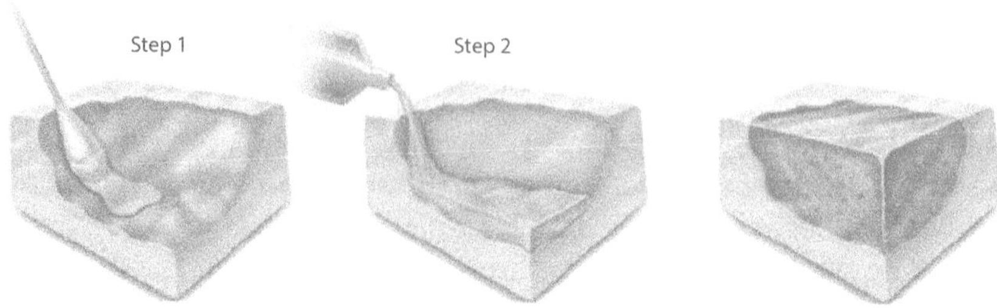

Figure 7.6 Bioadhesive tissue engineering [50].

and minimum irritation. Even fibrin glue is used in cardiac rehabilitation scaffold in integration with grafts [49].

7.7 Conclusion

In industrial applications, bioadhesives contain larger potentials to explore. In recent times, focus has been given in developing bioadhesives which are expected to result in products with high reliability and biocompatibility. In this view, some natural adhesives like mussels are becoming popular topics of research as far as the mechanism are concerned. Likewise, tissue adhesives are having potential to be used as key adhesives in haemostatic and wound healing applications while they are currently used as a standby adhesive. Even protein-based adhesives render better mechanical properties and biocompatibility when compared with all other adhesives and glues. They are currently under research along with the traditional polymeric materials for exploring the effectiveness in numerous applications. Fields like packaging, construction, food, and health need sustainability which could be easily achieved by the use of bioadhesives. Especially in medical field combination of bioadhesives with medicated internally used metallic elements are growing and widely used by surgeons, doctors, and patients. In technical, legal, medical, and other industrial applications achieving biocompatibility along with the performance of the adhesives with higher adhesion properties is a greater challenge. Considering all the facts development of bioadhesives, not only for bio-related applications but also for metallic applications are inevitable. This may open doors for the researchers working on metals, alloys, and composite materials to have a better view upon avoiding the failure of the elements using bioadhesives.

References

1. Zhu, W., Chuah, Y.J., Wang, D.-A., Bioadhesives for internal medical applications: A review. *Acta Biomater.*, 2018, https://doi.org/10.1016/j.actbio.2018.04.034.
2. Hemmila, V., Adamopoulos, S., Karlsson, O., Kumar, A., Development of sustainable bio-adhesives for engineered wood panels-A Review. *RSC Adv.*, 2017, https://pubs.rsc.org/en/content/articlelanding/2017/ra/c7ra06598a
3. Zhang, H., Zhao, T., Newland, B., Duffy, P., Annaidh, A.N., O'Cearbhaill, E.D., Wang, W., On-demand and negative-thermo-swelling tissue adhesive based on highly branched ambivalent PEG-catechol copolymers. *J. Mater. Chem. B*, 2015, https://pubs.rsc.org/en/content/articlelanding/2015/TB/C5TB00949A#!divAbstract
4. Roy, S.K. and Prabhakar, B., Bioadhesive polymeric platforms for transmucosal drug delivery systems - A review. *Trop. J. Pharm. Res.*, 2010, https://www.researchgate.net/publication/43561170_Bioadhesive_Polymeric_Platforms_for_Transmucosal_Drug_Delivery_Systems_-_a_Review
5. Wheat, J.C. and Wolf, J.S., Advances in Bioadhesives, Tissue Sealants, and Hemostatic Agents. *Urol. Clin. North Am.*, 2009, https://www.ncbi.nlm.nih.gov/pubmed/19406326
6. Guo, J., Wang, W., Hu, J., Xie, D., Gerhard, E., Nisic, M., Shan, D., Qian, G., Zheng, S., Yang, J., Synthesis and characterization of anti-bacterial and anti-fungal citrate-based mussel-inspired bioadhesives. *Biomaterials*, 2016, https://www.sciencedirect.com/science/article/pii/S0142961216000958
7. Forooshani, P.K. and Lee, B.P., Recent approaches in designing bioadhesive materials inspired by mussel adhesive protein. *J. Polym. Sci. Part A: Polym. Chem.*, 2017, https://onlinelibrary.wiley.com/doi/full/10.1002/pola.28368
8. Shaikh, R., Singh, T.R.R., Garland, M.J., Woolfson, A.D., Donnelly, R.F., Mucoadhesive drug delivery systems. *J. Pharm. Bioallied*, 2011, https://www.ncbi.nlm.nih.gov/pmc/articles/PMC3053525
9. Khanlari, S. and Dube, M.A., Bioadhesives: A review. *Macromol. React. Eng.*, 2013, https://www.onlinelibrary.wiley.com/doi/full/10.1002/mren.201300114
10. Vasir, J.K., Tambwekar, K., Garg, S., Bioadhesive microspheres as a controlled drug delivery system. *Int. J. Pharm.*, 2003, https://www.sciencedirect.com/science/article/pii/S0378517303000875
11. Ahadian, S., Rahal, R., Ramon-Azcon, J., Obregon, R., Hasan, A., Biomaterials in Tissue Engineering, in: *Tissue Engineering for Artificial Organs: Regenerative Medicine, Smart Diagnostics and Personalized Medicine*, 2016.
12. Wang, Z., Li, C., Xu, J., Wang, K., Lu, X., Zhang, H., Qu, S., Zhen, G., Ren, F., Bioadhesive microporous architectures by self-assembling polydopamine microcapsules for biomedical applications. *Chem. Mater.*, 2015, https://pubs.acs.org/doi/10.1021/cm5038765

13. Mehdizadeh, M. and Yang, J., Design Strategies and Applications of Tissue Bioadhesives. *Macromol. Biosci.*, 2013, https://onlinelibrary.wiley.com/doi/pdf/10.1002/mabi.201200332
14. Lacoste, C., El Hage, R., Bergeret, A., Corn, S., Lacroix, P., Sodium alginate adhesives as binders in wood fibers/textile waste fibers biocomposites for building insulation. *Carbohydr. Polym.*, 2018, https://www.sciencedirect.com/science/article/pii/S0144861717314248
15. Mohanty, A.K., Misra, M., Drzal, L.T., *Natural fibers, biopolymers, and biocomposites*, CRC Press, https://www.taylorfrancis.com/books/9780429211607
16. Bhushan, B., Tokachichu, D.R., Keener, M.T., Lee, S.C., Morphology and adhesion of biomolecules on silicon based surfaces. *Acta Biomater.*, 2005, https://www.sciencedirect.com/science/article/pii/S1742706105000462
17. Kong, X., Liu, G., Curtis, J.M., Characterization of canola oil based polyurethane wood adhesives. *Int. J. Adhes. Adhes.*, 2011, https://www.sciencedirect.com/science/article/pii/S0143749611000662
18. Mathias, J.D., Grédiac, M., Michaud, P., Bio-based adhesives, in: *Biopolymers and Biotech Admixtures for Eco-Efficient Construction Materials*, 2016.
19. Richter, K., Grunwald, I., von Byern, J., Bioadhesives, in: *Handbook of Adhesion Technology*, L. da Silva, A. Oechsner, R. Adams (Eds.), Springer, Cham, 2018, https://doi.org/10.1007/978-3-319-42087-5_53-2.
20. Schricker, S.R., Palacio, M.L.B., Bhushan, B., Review article: Designing nanostructured block copolymer surfaces to control protein adhesion. *Philos. Trans. R. Soc. A: Math. Phys. Eng. Sci.*, 2012, https://royalsocietypublishing.org/doi/10.1098/rsta.2011.0484
21. Brubaker, C.E. and Messersmith, P.B., The present and future of biologically inspired adhesive interfaces and materials. *Langmuir*, 2012, https://infoscience.epfl.ch/record/175744
22. Logan, R.M., The science of adhesion. *J. Acoust. Soc. Am.*, 2015, https://www.adhesives.org/adhesives-sealants/science-of-adhesion
23. Annabi, N., Yue, K., Tamayol, A., Khademhosseini, A., Elastic sealants for surgical applications. *Eur. J. Pharm. Biopharm.*, 2015, www.sciencedirect.com/science/article/pii/S0939641115002544
24. Lai, X.L., Deng, H.W., Yao, X.M., Li, M., Tan, W.J., Research on the improved fibrin adhesive in pterygium surgery. *Int. Eye Sci.*, 2013, 10.3980/j.issn.1672-5123.2013.02.32
25. Lu, Y.P., Yang, C.H., Yeh, J.A., Ho, F.H., Ou, Y.C., Chen, C.H., Lin, M.Y., Huang, K.S., Guidance of neural regeneration on the biomimetic nanostructured matrix. *Int. J. Pharm.*, 2014, https://www.sciencedirect.com/science/article/pii/S037851731300728X
26. Zhong, C., Gurry, T., Cheng, A.A., Downey, J., Deng, Z., Stultz, C.M., Lu, T.K., Self-assembling multi-component nanofibers for strong bioinspired underwater adhesives. *Nat. Nanotechnol.*, 2014, http://hdl.handle.net/1721.1/100833

27. Cha, H.J., Hwang, D.S., Lim, S., Development of bioadhesives from marine mussels. *Biotechnol. J.*, 2008, https://onlinelibrary.wiley.com/doi/abs/10.1002/biot.200700258
28. Arnold, M., Cavalcanti-Adam, E.A., Glass, R., Blummel, J., Eck, W., Kantlehner, M., Kessler, H., Spatz, J.P., Activation of integrin function by nanopatterned adhesive interfaces. *ChemPhysChem*, 2004, https://onlinelibrary.wiley.com/doi/abs/10.1002/cphc.200301014
29. Vaidyanathan, T.K. and Vaidyanathan, J., Recent advances in the theory and mechanism of adhesive resin bonding to dentin: A critical review. *J. Biomed. Mater. Res.-Part B Appl. Biomater.*, 2009, https://www.onlinelibrary.wiley.com/doi/full/10.1002/jbm.b.31253
30. Khang, D., Lu, J., Yao, C., Haberstroh, K.M., Webster, T.J., The role of nanometer and sub-micron surface features on vascular and bone cell adhesion on titanium. *Biomaterials*, 2008, https://www.sciencedirect.com/science/article/pii/S0142961207008988
31. Shekaran, A. and Garcia, A.J., Nanoscale engineering of extracellular matrix-mimetic bioadhesive surfaces and implants for tissue engineering. *BBA - Gen. Subjects*, 2011, https://www.ncbi.nlm.nih.gov/pubmed/20435097
32. Lu, J., Khang, D., Yao, C., Haberstroh, K.M., Webster, T.J., Increased vascular cell adhesion on titanium: The role of nano and sub-micron surface features, in: *8th World Biomaterials Congress 2008*, 2008.
33. Cooper, K., Gupta, A., Beaudoin, S., Simulation of the adhesion of particles to surfaces. *J. Colloid Interface Sci.*, 2001, https://www.sciencedirect.com/science/article/abs/pii/S002197970097276X
34. Forooshani, P.K. and Lee, B.P., Recent approaches in designing bioadhesive materials inspired by mussel adhesive protein. *J. Polym. Sci. Part A: Polym. Chem.*, 2016, https://onlinelibrary.wiley.com/doi/full/10.1002/pola.28368
35. Baldan, A., Adhesively-bonded joints and repairs in metallic alloys, polymers and composite materials: Adhesives, adhesion theories and surface pretreatment. *J. Mater. Sci.*, 2004, https://doi.org/10.1023/B:JMSC.0000007726.58758.e4
36. Marco, A., Elisabetta Ada, C.-A., Roman, G., Jacques, B., Wolfgang, E., Martin, K., Horst, K., Joachim, P.S., Activation of integrin function by nanopatterned adhesive interfaces. *ChemPhysChem*, 2004, https://onlinelibrary.wiley.com/doi/abs/10.1002/cphc.200301014
37. Perez, E. and Pincet, F., Bioadhesion, in: *Peyresq Lectures on Nonlinear Phenomena*, 2012, https://doi.org/10.1142/9789814440592_0006
38. Haiat, G., Phan Huy, M.C., Barthel, E., The adhesive contact of viscoelastic spheres. *J. Mech. Phys. Solids*, 2003, www.sciencedirect.com/science/article/pii/S0022509602000595
39. Kantibhai, P.T., Patel, K.S., Shah, S., Test methods of bioadhesive system. *World J. Pharm. Res.*, 7, 2, 267–278, 2018.

40. Blanco-Fuente, H., Anguiano-Igea, S., Otero-Espinar, F.J., Blanco-Méndez, J., *In-vitro* bioadhesion of carbopol hydrogels. *Int. J. Pharm.*, 1996, https://www.deepdyve.com/lp/elsevier/in-vitro-bioadhesion-of-carbopol-hydrogels-7YmUghKlst
41. Hannig, C. and Hannig, M., The oral cavity - A key system to understand substratum-dependent bioadhesion on solid surfaces in man'. *Clin. Oral Investig.*, 2009, https://link.springer.com/article/10.1007/s00784-008-0243-3
42. Wang, J., Tahir, M.N., Kappl, M., Tremel, W., Metz, N., Barz, M., Theato, P., Butt, H.J., Influence of binding-site density in wet bioadhesion. *Adv. Mater.*, 2008, https://doi.org/10.1002/adma.200801140
43. Chavanpatil, M.D., Jain, P., Chaudhari, S., Shear, R., Vavia, P.R., Novel sustained release, swellable and bioadhesive gastroretentive drug delivery system for ofloxacin. *Int. J. Pharm.*, 2006, https://www.sciencedirect.com/science/article/pii/S0378517306001839
44. Singh, I. and Paramjot, P., Bioadhesives in drug delivery, in: *Handbook of Adhesive Technology*, Third Edition, 2017.
45. He, Y., Peng, Y., Li, Z., Ma, J., Zhang, X., Liu, K., Wang, W., Jiang, L., Bio-inspired multifunctional metallic glass. *Sci. China Chem.*, 2016, https://link.springer.com/article/10.1007/s11426-015-5496-5
46. Rathi, S., Saka, R., Domb, A.J., Khan, W., Protein-based bioadhesives and bioglues. *Polym. Adv. Technol.*, 30, 217–234, 2019, https://doi.org/10.1002/pat.4465.
47. Kumar, K., Dhawan, N., Sharma, H., Vaidya, S., Vaidya, B., Bioadhesive polymers: Novel tool for drug delivery. *Artif. Cells Nanomed. Biotechnol.*, 2014, https://doi.org/10.3109/21691401.2013.815194
48. Xu, J., Strandman, S., Zhu, J.X.X., Barralet, J., Cerruti, M., Genipin-crosslinked catechol-chitosan mucoadhesive hydrogels for buccal drug delivery. *Biomaterials*, 2015, https://www.ncbi.nlm.nih.gov/pubmed/25453967
49. Haag, A.P., Geesey, G.G., Mittleman, M.W., Bacterially derived wood adhesive. *Int. J. Adhes. Adhes.*, 2006, https://www.sciencedirect.com/science/article/abs/pii/S0143749605000503
50. Moubarik, A., Allal, A., Pizzi, A., Charrier, F., Charrier, B., Preparation and mechanical characterization of particleboard made from maritime pine and glued with bio-adhesives based on cornstarch and tannins. *Maderas: Cienc. Tecnol.*, 2010, https://scielo.conicyt.cl/scielo.php?script=sci_arttext&pid=S0718-221X2010000300004

8

Polysaccharide-Based Adhesives

Asad Ali[1], Kanwal Rehman[2], Humaira Majeed[1], Muhammad Fiaz Khalid[1] and Muhammad Sajid Hamid Akash[1]*

[1]*Department of Pharmaceutical Chemistry, Government College University Faisalabad, Pakistan*
[2]*Department of Pharmacy, University of Agriculture, Faisalabad, Pakistan*

Abstract

Synthetic adhesives are known for their toxic and unhealthy impact on natural environment. The development of natural adhesives based on renewable resources is dire necessity of this age. Their importance is growing gradually as the commercial market is motivated to emphasis on the benefits of green adhesives. Natural adhesives particularly emerged from polysaccharides have number of good features like biodegradability, biocompatibility, antimicrobial activity, non-toxicity, and low cost to keep a good pace with synthetic adhesives. In this regard, chitosan has recognized as a potential bioadhesive used in tissue engineering and muco-adhesive drug delivery system. Similarly, the bioadhesives have also been marked effective for the treatment of optic and surgical wounds in area of ophthalmology. In addition, the adhesive nature of polysaccharides increases the holding capacity of foliage to accumulate active substances. Herbicide and pesticide sprays, xanthan gum, etc., help to wet the leaves so that adhesion may establish and active ingredients may be able to perform their activity. Moreover, they have demonstrated themselves as a potential excipient in pharmaceutical field. By the same token, Cigarette making industries are also utilizing starch-based adhesives for manufacturing rod and plug of a cigarette. Interestingly, their role as a skin cleanser has also been reported. Last but not least, they have also found their extensive applications in food industry. This particular review explains different natural resources of adhesives, their biotechnological development methods, current applications, and a brief tale of different mechanical tests to evaluate their efficiency.

Keywords: Bio-adhesive, adherend, adhesion strength, films, shear

**Corresponding author*: sajidakash@gcuf.edu.pk

Inamuddin, Rajender Boddula, Mohd Imran Ahamed and Abdullah M. Asiri (eds.) Green Adhesives: Preparation, Properties and Applications, (165–180) © 2020 Scrivener Publishing LLC

8.1 Introduction

Polysaccharides are composed of polymeric carbohydrates. They are composed of repeating units of monosaccharides connected with glycosidic linkage with numerous hydroxyl groups [1]. They are abundantly found in all plants [2], outer shell of aquatic animals [3], and few microbes [4]. It is interesting to know that minimum 90% of carbohydrates occurred in nature are polysaccharides. They include 70% of dry weight of total biomass [5]. Therefore, they are easily available at cheapest prices [6]. Polysaccharides especially from plants are playing their integral role in daily life of a man by providing essential nutrients to health and benefitting textile, paper, petrochemical, and pharmaceutical industry. Moreover, their role as bio-adhesive enjoys supreme place in commercial market nowadays because of their biocompatibility, biodegradability, non-toxicity, antimicrobial activity, and low cost. As the polysaccharides have natural and renewable bioresources so they have replaced a number of non-renewable adhesives for instance poly (vinylacetate), epoxy, phenol-formaldehyde, and polyurethane. Furthermore, the toxic chemical substances of synthetic adhesives are not only a serious threat to health but also a big challenge to environmental protection agencies of world. The development of green bioadhesives is the prime goal of these agencies. They are also encouraging synthetic industry to resolve these environmental and health safety related issues and manufacture environmental-friendly biological adhesives. According to Adhesives and Sealants Market Size, Share & Trends Analysis Report (2019–2024), the adhesives and salants market size is forecasted to be USD 58.9 billion in 2019 and expected to reach USD 73.8 billion by 2024, at a compound annual growth rate (CAGR) of 4.6%, from 2019 to 2024. Moreover, The Adhesives and Sealants (ASC) council is also financially supporting to manufacturing companies for developing bio-based alternatives to tackle environmental hazards [7]. Bioadhesives based on their structures are classified into proteins, polyphenols, polythioesters, polysaccharides, polyisoprenoids polyoxoesters, nucleic acids, and polyesters. Among them, polysaccharides being involved in bio-film formation and natural mediator between surfaces are termed as potential bioadhesive. Water-based, foamed polysaccharides are used for bonding of such substances having fiber materials such as paper, paper board, card board, etc. [8].

8.2 Cellulose-Derived Adhesive

Cellulose is present in cell wall of all plants as a main structural component. It is also a main participant in cell wall composition of microbes and few animals like tunicates. In addition, annual production of cellulose is estimated about 10 tons to make it one of the most striking renewable resources [9]. The repeating units of cellulose are termed as cellobiose. A number of hydroxyl groups present on cellulose molecule allow it to form variety of derivatives. Amazingly, these derivatives give significant adhesive property to cellulose. On the other hand, hydrogen bonds present on cellulose don't break easily so the cellulose molecules are not readily soluble in many solvents. Hence, the role of cellulose itself as an adhesive comes to an end. To overcome this problem, the hydroxyl ends of cellulose undergo esterification and etherification reactions to produce such derivatives that can easily soluble in many common solvents. Cellulose modified adhesives developed through esterification and etherificaion are being discussed in following lines.

8.2.1 Esterification

Esterification is a chemical reaction in which alcohols and acids are main reactants that are known for the synthesis of esters. Common starch derived esters which play their role as potential adhesives includes, cellulose acetate, cellulose nitrate, and cellulose acetate butyrate.

8.2.1.1 Cellulose Nitrate

The procedure of esterification, for the first time, was found by a scientist named as Schonbein. He produced cellulose nitrate by setting up a mix reaction among cellulose, sulfuric acid, and nitric acid. Cellulose nitrate obtained from this reaction was flammable and have higher degree of substitution (DS). Being flammable, its main use was restricted to smokeless gunpowder. Afterwards, the efforts were made to reduce DS to make its application in other fields as well. The preparation of cellulose nitrate is depicted in (Figure 8.1a) where there is a chemical reaction between cellulose and nitric acid catalyzed by sulfuric acid giving cellulose nitrate and excess of water. Finally, boiling is carried out to take away sulfate groups to

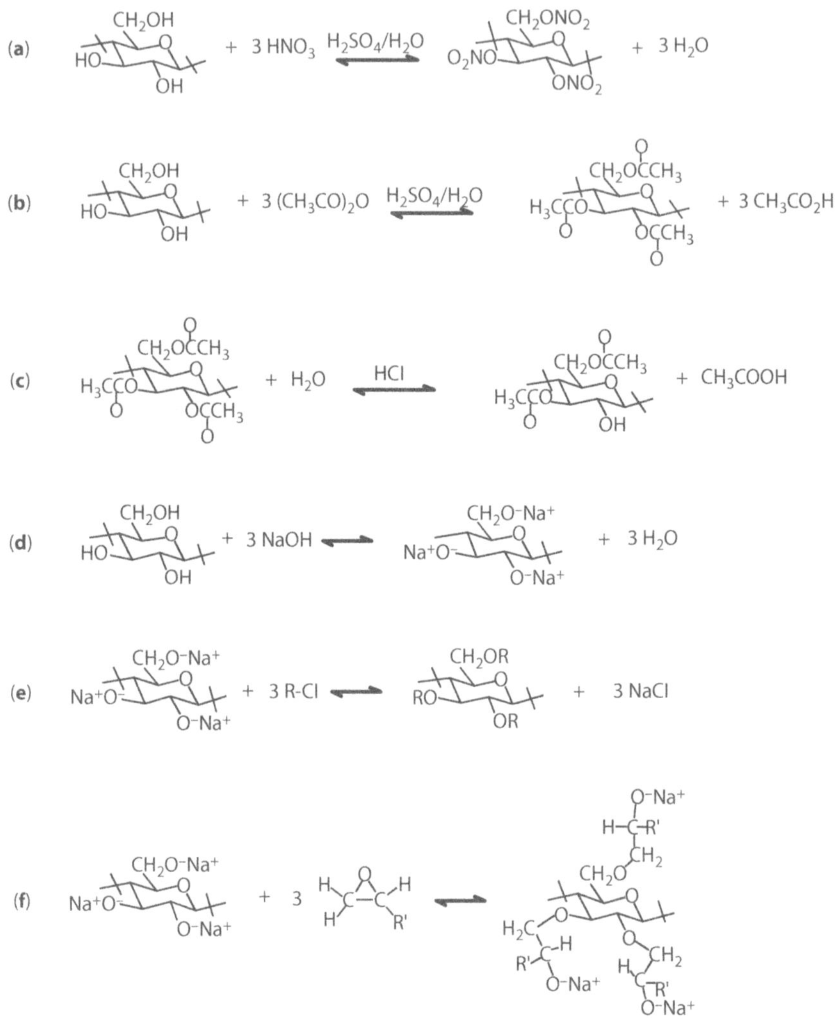

Figure 8.1 Major reaction of cellulose: (a) nitration by nitric acid; (b) esterification by acetic anhydride; (c) hydrolysis of tertiary cellulose acetate by hydrochloric acid; (d) deprotonation of cellulose, the first step in etherification; (e) etherification of sodium cellulose by an alkyl halide; (f) etherification of sodium cellulose by an epoxide. Adopted from (Baumann *et al.*, 1994).

maintain stability of cellulose nitrate. Cellulose nitrate commonly used as household cement to form durable, wet-resistant, and clear films. It gives paint films when a suitable plasticizer, pigment, or resin is added to cellulose nitrate [10].

8.2.1.2 Cellulose Acetate

Its use as an adhesive is not much prominent. Being more heat resistant and less water resistant as compare to cellulose nitrate it becomes brittle within time. It is produced by a chemical reaction of acetic anhydride and cellulose, catalyzed by sulfuric acid (Figure 8.1b). Cellulose acetate produced has greater DS, not suitable, to act as a potential bioadhesive. So, the cellulose acetate is further hydrolyzed by hydrochloric acid to reduce its DS (Figure 8.1c).

8.2.1.3 Cellulose Acetate Butyrate

Cellulose acetate butyrate is more water proof and greatly soluble in many solvents as compare to cellulose acetate. It is used in paper industry to avoid paper from staining. There are two common reactions reported for preparation of cellulose acetate butyrate. In first reaction, cellulose nitrate is obtained by anhydrides with suitable proportions in the presence of sulfuric acid. In 2nd reaction, it is produced in a chemical reaction between propionic acid or butyric acid and acetic anhydride, catalyzed by sulfuric acid.

8.2.2 Etherification

This is a process in which cellulose is treated with sodium hydroxide (Figure 8.1d) to obtain sodium cellulose. Then, the chemical reaction between alkyl halide and sodium cellulose gives sodium halide along with cellulose ether (Figure 8.1e). Similarly, sodium cellulose is also etherified with epoxide (Figure 8.1f).

8.2.2.1 Methyl Cellulose

A chemical substance named as methyl chloride reacts with cellulose to give methyl cellulose. Methyl cellulose is used as lining adhesive in a technique named as remoistenable backing. The development of its films is contraction and distortion free. The restoring feature of films after exposing with moisture makes them superior to films of starch paste. Their aging feature is applied in conservation treatments like sizing, consolidation, and supporting paper objects [11]. It is also used for paper coating and sizing to impart grease resistance in ceramics as a binder, as a non-staining paste for wallpaper, and as an adhesives for leather drying [12].

8.2.2.2 Ethyl Cellulose

Biofilms produced by ethyl cellulose in wet conditions do not absorb moisture and show more resistance to chemical break down. Its synthesis is very simple as methyl cellulose. It is obtained by chemical reaction between cellulose and ethyl chloride. Ethyl cellulose and chitosan both combined together may be used as an adhesive in mucoadhesive drug delivery system. It has also become clear in finding its application in new mucoadhesive bilayered devices made of chitosan as adhesive and ethyl cellulose as an impermeable backing material [13].

8.2.2.3 Carboxymethyl Cellulose

Sodium salt of carboxymethyl cellulose is water soluble used as detergents to remove stains. It is widely used as thickener and being non toxic has great importance in food industry. Novel modification of starch/carboxymethyl cellulose produces composite films by using casting method. The addition of carboxymethyl cellulose gives blend films with optimum tensile strength. The resultant films don't break easily as compare to pure starch films [14].

8.3 Starch-Derived Adhesives

Starch is a storage polysaccharide comprising repeating units of glucose. Amylose as linear and amylopectin as branched is the simplest form of starch. Its application as material binder in 1985 counts more than 1.6 billion kilograms that shows its extensive use. Similarly, there were almost a half-billion pounds of pressure sensitive adhesives sold in the same year [15]. So far, the commercial market is getting enlarged and making starch the most striking adhesive. Additionally, dextrins are obtained by the hydrolysis of dry-roasted starch. Dextrins are classified into three groups. These are white dextrin, canary, or yellow dextrin and British gums. They are different from one another on the basis of roasting time, temperature and the quantity of catalyst used. Besides, generally, some additives are added to starch derivatives for increasing the viscosity and thickness of suspensions. They are also used to avoid from brittleness of the glue line. Additives like plasticizers (urea, sodium nitrate, salicylic acid, and formaldehyde), colloid stabilizer (soaps and sodium chloride), and fillers (clay and bentonite) are widely used. Some preservatives (formaldehyde), defoamers, bleaches (sodium bisulfite), and organic solvents are also used to kill microbes, stop foaming, remove discoloration of glue, and increase bonding tendency, respectively. Water resistance and bonding strength was increased remarkably when additives and isocyanatess were mixed with starch adhesives [16]. Starch and dextrin are used in paper and textile industry as binders and sizing materials. Moreover, they have won special attention in manufacturing paper bags and sacks, laminating boards, and walls and carton sealing. Starch itself is unable to act as adhesive because of branched and chain alignment of amylose and amylopectin. The branches must be opened to make starch a suitable candidate as an adhesive after applying some methods like alkali treatment, acid treatment, heating, and oxidation. These are being explained briefly in following headings.

8.3.1 Alkali Treatment

Gelation occurs at room temperature when sufficient amount of alkali is added. When coir fibers were mixed in starch/ethylene vinyl alcohol copolymers (EVOH) blends after mercerized with alkali. The resultant fibers had tensile strength increased by 53% approximately as compare to untreated fibers [17].

8.3.2 Acid Treatment

The starches are heated at 49°C to 54°C with sufficient amount of mineral acid followed by suitable base is added for neutralization. In addition, a stable adhesive with high viscosity and thickness is synthesized firstly by treating starch with acid or alpha-amylase followed by further reactions with beta-amylase. Resultant product gives good characteristics like better adhesion, bond strength, and stability [18].

8.3.3 Heating

Firstly, the starch granules are heated until the thickening of suspension. For starches in pure water, gelation occurs between 57°C and 72°C. And the temperature at which gelation occurs is called as gelation temperature. Finally, this gelation disrupts the crystalline nature of starch. Moreover, starch-based adhesives are prepared by hydrolyzing starch with polyvinyl alcohol under alkaline and acidic conditions at different temperatures. The bond strength of starch reached to maximum and compatibility between starch and poly(vinyl acetate) also increased as the temperature increases [19].

8.3.4 Oxidation

Starch is oxidized by treating starch suspension at pH 8 to 10 with hypochlorite solution until the thickening occurs. The peak viscosities of oxidized rice and corn starches are more viscous than unoxidized starch and corn at 0.8% NaOCl [20]. Oxidized starch is more stable, viscous, and has high degree of adhesion. Hypochlorite is considered one of the most efficacious oxidant.

8.4 Natural Gums Derived-Adhesives

When gums are added in aqueous medium, they produce viscous suspensions. Natural gums include varieties like guar gum, locust bean gum, xanthan gum, gum tragacanth, tamarind gum, gum arabic, gellan gum, gum karaya, gum ghatti, algal gums (gum agar and alginates). *Terminalia bellerica* (Bahera), a natural gum, when used as an additive in water-based natural rubber adhesive formulations, the lap shear strength of joints with wood substrates increased with the increase of natural gum concentration up to 80% loading. Similarly, the peel strength was also increased with addition of the gum [21]. Natural gums have found vast application in paper, textile, food, pharmaceutical industry, heavy ion separations, laundry, and household products because of their potential to make gel and also having stabilizing character. Despite of many limitations, these adhesives are used to bind wood-wood and many common interfaces. Moreover, Natural gums are not only marked as environment friendly but also used as adsorbent to separate heavy metal ion from industrial sewage [22].

8.5 Fermentation-Based Adhesives

One of the studies reveal that the fermentation of alfalfa (*Medicago sativa L.*) gives comprising fibers, fermented residues (FR), and glycocalyx in the presence of *ruminococcus albus or clostridium thermocellum*. Then, these microbial-fermented plant fibers are replaced with synthetic part of adhesives. These fermented biomass co-adhesives work well under wet conditions and exhibit good shear strength. The presence of glycocalyces further improves adhesion and binding ability. In addition, cassava is a tropical crop that is serving about 700 million people by providing 200–1000 calories in a day. For achieving a high quality and stable adhesive, cassava was kept under fermentation process. Amazingly, cassava starch with high HCN (hydrocyanic acid) content and low pH gave high quality and stable adhesives. But such fermented cassava was only confined to industries being toxic in nature. However, if adhesive for domestic application is required, then low HCN starch with salts and non-toxic oxidizing agents such as ferrous sulphate, aluminum chloride, sodium hypochlorite, and acetic acid are mixed to extend the shelf life [23].

8.6 Enzyme Cross-Linked-Based Adhesives

This is special adhesive system in which chitosan, phenolic compounds, and enzyme laccase are main three participants. Here, there is formation of covalent bonding between the amino group of chitosan and the quinine group of phenolic compound by Schiff base mechanism. Laccase plays a role as oxidizing the phenolic compounds. Resultantly, such bio adhesives are produced having durable binding strength. Similarly, a lignin-based adhesive was also produced by replacing the synthetic latex in wool floor covering by a laccase-enhanser system followed by a phenolation step where natural phenolic compounds were copolymerized with lignin in order to improve its content of quinone structures reactive towards wool. The synthesized adhesive was flexible, having increased bonding strength [24].

8.7 Micro-Biopolysaccharide-Based Adhesives

Extracellular polysaccharides produced by microorganisms are also considered good candidate as bio adhesives. Although, these types of adhesives do not work well as compare to wood adhesives present in the market, but nowadays, they have proved themselves as a good competitor for bonding applications. Besides, microbial fermentation process was applied on a non-petrochemical feedstock to synthesize a stable and more resistant adhesive. In this study, even though, resultant adhesive was susceptible to water but proved as strongly resistant to many solvents. The mean value of shear strength was 819 psi by using anodized aluminum [25].

8.8 Mechanism of Adhesion

How an adhesive works to bind two surfaces is very interesting and involves number of theories to describe adhesion phenomenon. In this effort, modern adsorption and mechanical theories of adhesion were introduced by McBain and Hopkins in 1925 [26]. According to them, the adhesion is governed by binding of adhesive in cavities or pores of the surface. Wet ability factor is considered favorable in this regard. Van der Leeden and Frens proposed that binding of adhesive was based on the shape of the surfaces and rheological properties [27]. Whenever the adhesive could not be able to penetrate into cavities or rough surfaces it tends to form air bubbles weakening the penetration between adhesive and adherend. The mechanical interlocking theory explains not only the porous or rough substrates such as wood and paper but also informs about some metallic substrates such as anodized aluminum [28]. John Venables, an eminent scientist who developed scanning electron microscope to observe isomeric drawings showing all characteristics of oxidized metal surfaces. Actually, his work emphasizes on anodized layer of aluminum and titanium. In this spell, Keller, Hunter, and Robinson concluded the structure of porous oxide films on aluminum. But, according to Venables, porous surfaces of these anodized layers are micromechanical keying of the adhesives [29]. The adsorption theory also explains the mechanism of adhesion. According to this theory there are intermolecular forces present between the molecules of adhesive and adherend [30]. In addition, the electrostatic theory tells about the possible mutual electron sharing between adhesive and adherend. The diffusion or inter diffusion theory explained by Voyutskii deals with those polymers which are mutually miscible and compatible [31]. Fick's law of diffusion gains attention here. The thermodynamic theory also emphasizes on the direct measure of forces involved in making contact angle between the adhesive and the adherend at the thermodynamical equilibrium. Current theories of adhesion have been reviewed with particular regard to the problems of adhesive dentistry. Although, Adhesive material is currently available to be applied at etched enamel yet suitable adhesive for tooth substances is demanding attention [32].

8.9 Tests for Adhesion Strength

There is a plethora of tests available to determine the strength of adhesives but only a few are capable of providing quantitative data that can readily be used in design calculations. These test methods are specially made to place loads creating stresses on adhesive joints. The test methods to determine the strength of an adhesive include tension, shear, cleavage, compression, peel, and stresses as shown in (Figure 8.2a). Their main function is to generate shear stress for evaluating strength of an adhesive.

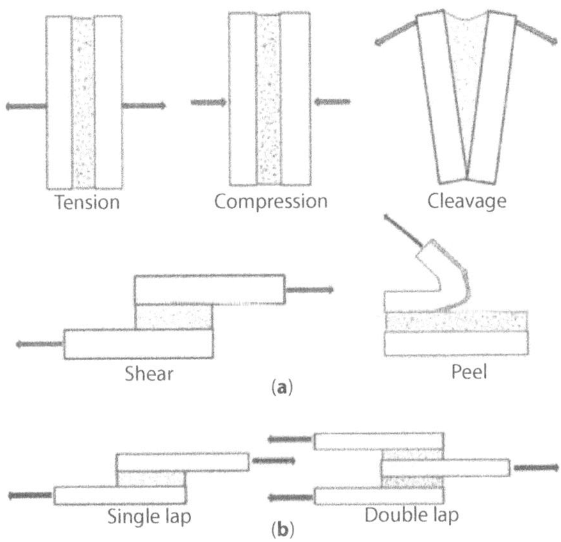

Figure 8.2 (a) Various modes of mechanical testing. (b) Single lap an double lap joints. Adopted from (Patel *et al.*, 2013).

Mechanical tests are necessary for determining significant properties like deformability, tensile and strength, and brittleness [33]. Mechanical tests include some of the most familiar adhesive joint tests such as the thin lap shear, the double lap shear, strap joint, and thick adherend shear tests. These lap joints are extensively used to determine the performance ability of an adhesive because they offer more reliable strength measurement as shown in (Figure 8.2b). Thin lap shear or single lap joint tests have been identified for some limitations for the accurate results as the stresses are loaded at the ends of the overlap. This eccentric loading produces bending of adherends and rotation of bonding region [34]. This problem was resolved by developing a new mechanical test known as double-lap joints as shown in (Figure 8.3) that represents a scheme of symmetrical double lap joint where the central adherend is twice the thickness of the outer adherends. Moreover, single-lap tests are replaced by double tap tests as the former produces coupling effects. While double-lap tests measure the shear capacity of adhesives with less peel effects. In addition to this, thin adhesive films are used in packaging, coating or for advertising. Once a film is adhered to a substrate, flaps can be detached by tearing and peeling. This triangular shape of the tear determines the mechanical characterization of thin adhesive films, nanofilms deposited on substrates or fruit skin [35]. So, tearing is a good tool to evaluate mechanical characterization of thin adhesive films.

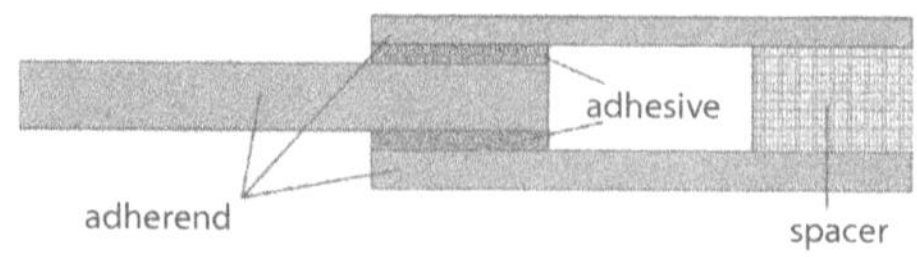

Figure 8.3 Scheme of lap joint test. Adopted from Duncan and Crocker 2001.

8.10 Applications

8.10.1 Biomedical Applications

This is generally said that polysaccharide adhesives are only confined to binding adherends. But, recent studies have reported that chitosan-based bio adhesives have got significant attention in pharmaceutical and biomedical field. They have found their application in tissue engineering and wound dressing [36]. Their mucoadhesive property along with water resistance and release characteristics has made them attractive for mucoadhesive drug delivery system. In addition to this, poly glycolic acid-chistosan glue has been synthesizes by the deacetylation of chitosan. Interestingly, this new modified glue works better than fibrin glue. Biochemical studies have revealed its clinical application because of its outstanding biocompatibility [37]. Furthermore, polysaccharides cross-linked with dextrans are also drawing attention in wound treatment. Cationic polysaccharides have their extensive use in biomedical field as well. They are equally important in internal surgical practices and have shown antibacterial activity. Bioadhesive polymers like sodium carboxymethylcellulose (Na-CMC), hydroxypropyl methylcellulose (HPMC), levan, xanthan is being used in drug delivery systems. Moreover, mucoadhesive drug delivery system also increases bioavailability and local therapeutic effect of drugs by improving the residence time at the site of application [38]. By the same token, polysaccharide-based adhesives have also drawn attention for surgical and optic wounds in the field of ophthalmology. Modern adhesives emerged from protein tissue solder and hydrogels are being used nowadays. These new adhesive systems are more effective than already available sutures in market [39].

8.10.2 Food Stuffs Applications

Dextrin-based food grade adhesives with little amount of carboxymethylcellulose, xanthan, or its combinations are used to keep spiced and flavored coatings intact in manufacturing process of oil roasting of nut meat products. The coatings of dextrin-based food grade adhesives having no xanthan, carboxy methyl cellulose, or its combinations are vulnerable to oil roasting [40].

The dried sap of gum arabic with immense commercial value is considered now a global commodity having good features like emulsifier, stabilizer, binder, and shelf-life enhancer. Hence, it is extensively used in many foods [41].

8.10.3 Pharmaceutical Applications

Polysaccharides including starch and its derivatives have found their applications as binders, diluents, glidants, fillers, and disintegrants. Starch being low compatible and enzymatically degradeable was unable to tackle structural integrity and controlled release issues. To overcome this problem, various physical, chemical, and enzymatic hydrolysis methods were applied to produce such derivatives which are sufficient to make starch a good pharmaceutical excipient. Starch modifiers like retrograded starch, carboxymethyl starch, grafted starch, pregelatinized starch, starch acetate, cross-linked starch are used in oral tablets to control dug release [42].

8.10.4 Agricultural Applications

Moreover, their application to agricultural field can't be underestimated. The adhesive properties of polysaccharides provide holding capacity to foliage so that active substances remain stick to foliage surface. Herbicide and pesticide sprays, starch, scleroglucan, xanthan gum, and carboxymethylcellulose help to wet the leaves so that adhesion may establish and active ingredients may be able to perform their activity in an effective way [9].

8.10.5 Cigarette Manufacturing

Starches prepared at high temperature and pressure is used in the manufacturing of cigarettes. Cigarettes are composed of a cigarette rod and a filter or plug. A cigarette formation is incomplete without use of adhesives. Side seam and lap seal adhesives are attached to paper tube of cigarette rod. Centerline adhesive is attached to center of paper embedded in filter or filter plug and overlap adhesive forms filter plug. The filter or plug is then attached to the cigarette rod by wrapping another paper (tipping paper) which contains an adhesive (tipping adhesive) around the plug and overlapping it with an aligned rod [43].

8.10.6 Skin Cleansing Applications

Adhesives are also reported for cleansing of skin. Keratotic plugs and dead epidermal cells are known for pluging the skin pores. Keratotic plugs and dirt from skin pores cannot be removed by traditional cleansers and detergents. The only way to remove this debris is the application of cleansing adhesives. Additionally, the use of synthetic cationic polymers containing salt forming groups can remove these plugs but natural products are dire necessity now. The problem may arise for skin if adhesives are not properly removed from skin [44].

8.11 Conclusion

Polysaccharides are present in large amount in nature. That's why they are easily available at cheapest price. Their natural and renewable resources have made them attractive for commercial market in whole world. In near future, the scarcity of petrochemical resources will magnify their importance and divert the attention of industries towards renewable resources. To this day, their role as bioadhesive enjoys supreme position because of their biocompatibility, biodegradability, non-toxicity, antimicrobial activity, and low cost. The manufacturing of green adhesives will safeguard this earth from polluted and toxic chemicals. Although bioadhesives are fulfilling the demand of commercial market to some degree but the need of hour is to develop such structural modified polysaccharides which can improve wet strength of bonds and thermal properties. In this regard, cross-linking of polysaccharides with organic or inorganic modifiers, recombinant technology, and combination of superior additives with an actual adhesive may produce constructive results.

References

1. Norström, E. *et al.*, Gum dispersions as environmentally friendly wood adhesives. *Ind. Crops Prod.*, 52, 736–744, 2014.
2. Burton, R.A. and Fincher, G.B., Plant cell wall engineering: Applications in biofuel production and improved human health. *Curr. Opin. Biotechnol.*, 26, 79–84, 2014.
3. Mathur, N.K. and Narang, C.K., Chitin and chitosan, versatile polysaccharides from marine animals. *J. Chem. Educ.*, 67, 11, 938, 1990.
4. Bertocchi, C. *et al.*, Polysaccharides from cyanobacteria. *Carbohydr. Polym.*, 12, 2, 127–153, 1990.
5. BeMiller, J.N., Polysaccharides: Occurrence, significance, and properties, in: *Glycoscience: Chemistry and chemical biology*, pp. 1413–1435, 2008.
6. Vroman, I. and Tighzert, L., Biodegradable polymers. *Materials*, 2, 2, 307–344, 2009.

7. Adhesives & Sealants Market by Adhesive Formulating Technology (Water-Based, Solvent-Based, Hot-Melt, Reactive), Sealant Resin Type (Silicone, Polyurethane, Plastisol, Emulsion, Butyl, Polysulfide), Application, Region - Global Forecast to 2024, Sep 2019 October 12, 2019. Available from: https://www.marketsandmarkets.com/Market-Reports/adhesive-sealants-market-421.html.
8. Wastyn M. M., Kozich M., Seidl B., inventors; Agrana Staerke GmbH, assignee. Polysaccharide Adhesive. United States patent application US 16/060,848. 2019.
9. Patel, A.K., Mathias, J.-D., Michaud, P., Polysaccharides as adhesives. *Rev. Adhes. Adhes.*, 1, 3, 312–345, 2013.
10. Kawakami, M. *et al.*, Gas permeabilities of cellulose nitrate/poly (ethylene glycol) blend membranes. *J. Appl. Polym. Sci.*, 27, 7, 2387–2393, 1982.
11. Brückle, I., Update: Remoistenable lining with methyl cellulose adhesive preparation, in: *The Book and Paper Group Annual*, vol. 15, pp. 25–26, 1996.
12. Baumann, M.G. and Conner, A.H., Carbohydrate polymers as adhesives, in: *Handbook of adhesive technology*, pp. 299–313, 1994.
13. Remunan-Lopez, C. *et al.*, Design and evaluation of chitosan/ethylcellulose mucoadhesive bilayered devices for buccal drug delivery. *J. Controlled Release*, 55, 2–3, 143–152, 1998.
14. Ghanbarzadeh, B., Almasi, H., Entezami, A.A., Physical properties of edible modified starch/carboxymethyl cellulose films. *Innovative Food Sci. Emerg. Technol.*, 11, 4, 697–702, 2010.
15. Kennedy, H.M., Starch-and dextrin-based adhesives, in: *Adhesives from renewable resources*, pp. 326–336, 1989.
16. Wang, Z. *et al.*, Bonding strength and water resistance of starch-based wood adhesive improved by silica nanoparticles. *Carbohydr. Polym.*, 86, 1, 72–76, 2011.
17. Nadiha, M.N. *et al.*, Comparative susceptibilities of sago, potato and corn starches to alkali treatment. *Food Chem.*, 121, 4, 1053–1059, 2010.
18. Henley, M.J., Stable high viscosity starch based adhesive and method of preparation, 1998, Google Patents.
19. Yu, H. *et al.*, Effects of treatment temperature on properties of starch-based adhesives. *BioResources*, 10, 2, 3520–3530, 2015.
20. Kuakpetoon, D. and Wang, Y.J., Characterization of different starches oxidized by hypochlorite. *Starch-Stärke*, 53, 5, 211–218, 2001.
21. Saha, S. *et al.*, Studies of Terminalia bellerica (Bahera), a natural gum, as an additive in a water-based adhesive composition. *J. Adhes. Sci. Technol.*, 19, 15, 1349–1361, 2005.
22. Parija, S., Misra, M., Mohanty, A., Studies of natural gum adhesive extracts: An overview. *J. Macromol. Sci. Part C: Polym. Rev.*, 41, 3, 175–197, 2001.
23. Oghenejoboh, K., Effects of starch fermentation on the shelf-life of cassava starch based adhesive. *Br. Biotechnol. J.*, 2, 4, 257, 2012.

24. Aracri, E., Blanco, C.D., Tzanov, T., An enzymatic approach to develop a lignin-based adhesive for wool floor coverings. *Green Chem.*, 16, 5, 2597–2603, 2014.
25. Combie, J. *et al.*, Adhesive produced by microorganisms, in: *ACS symposium series*, ACS Publications, 2004.
26. Packham, D., Some contributions of surface analysis to the development of adhesion theories. *J. Adhes.*, 84, 3, 240–255, 2008.
27. van der Leeden, M.C. and Frens, G., Surface properties of plastic materials in relation to their adhering performance. *Adv. Eng. Mater.*, 4, 5, 280–289, 2002.
28. Fourche, G., An overview of the basic aspects of polymer adhesion. Part I: Fundamentals. *Polym. Eng. Sci.*, 35, 12, 957–967, 1995.
29. Packham, D., The Mechanical Theory of Adhesion—Changing Perceptions 1925–1991. *J. Adhes.*, 39, 2–3, 137–144, 1992.
30. Packham, D.E., Theories of fundamental adhesion, in: *Handbook of adhesion technology*, pp. 11–41, 2018.
31. Voyutski, S., The diffusion theory of adhesion. *Rubber Chem. Technol.*, 33, 3, 748–756, 1960.
32. Tay, W. and Von Fraunhofer, J., Adhesion in dentistry. *Surf. Technol.*, 7, 2, 157–163, 1978.
33. McBain, J. and Lee, W., Adhesives and adhesion mechanical properties of films of adhesives. *Ind. Eng. Chem.*, 19, 9, 1005–1008, 1927.
34. Duncan B., and Crocker L. *Review of tests for adhesion strength.* National Physical Laboratory; 2001.
35. Hamm, E. *et al.*, Tearing as a test for mechanical characterization of thin adhesive films. *Nat. Mater.*, 7, 5, 386, 2008.
36. Patel, A.K., Chitosan: Emergence as potent candidate for green adhesive market. *Biochem. Eng. J.*, 102, 74–81, 2015.
37. Hoffmann, B. *et al.*, Characterisation of a new bioadhesive system based on polysaccharides with the potential to be used as bone glue. *J. Mater. Sci.: Mater. Med.*, 20, 10, 2001–2009, 2009.
38. Kumar, K. *et al.*, Bioadhesive polymers: Novel tool for drug delivery. *Artif. Cells Nanomed. Biotechnol.*, 42, 4, 274–283, 2014.
39. Oelker, A.M. and Grinstaff, M.W., Ophthalmic adhesives: A materials chemistry perspective. *J. Mater. Chem.*, 18, 22, 2521–2536, 2008.
40. Morris, C.A., Dextrin-based food-grade adhesive including xanthan or carboxymethylcellulose or mixtures thereof, 1991, Google Patents.
41. Patel, S. and Goyal, A., Applications of natural polymer gum arabic: A review. *Int. J. Food Prop.*, 18, 5, 986–998, 2015.
42. Hong, Y., Liu, G., Gu, Z., Recent advances of starch-based excipients used in extended-release tablets: A review. *Drug Deliv.*, 23, 1, 12–20, 2016.
43. Mooney, N.T. *et al.*, Starch based natural adhesives used in cigarette manufacture, 1992, Google Patents.
44. Howard, D.L. *et al.*, Starch based adhesives for skin cleaning tape, 2001, Google Patents.

9

Wound Healing Adhesives

Humaira Majeed[1], Kanwal Rehman[2], Asad Ali[1], Muhammad Fiaz Khalid[1] and Muhammad Sajid Hamid Akash[1]*

[1]Department of Pharmaceutical Chemistry, Government College University Faisalabad, Pakistan
[2]Department of Pharmacy, University of Agriculture, Faisalabad, Pakistan

Abstract
Wound healing adhesives have gained amassed interest for wound healing after surgical incisions, endoscopic processes, dental procedures, ocular surgeries in both corneal and retinal, cataract, pancreatectomy, gastric incisions, inguinal hernia repair, variceal bleeding, knee arthroplasty, cardiac surgeries, diabetic foot ulcers, accidental lacerations, and burns. Use of wound healing adhesives is of clinical interest due to improved patient comfort, cost effectiveness, less skill required for its application, no need of local anesthetic, ease in application, and less tissue damage during sealing wounds. There is great need of searching wound healing adhesives having properties of biocompatibility, biodegradability, no toxicity, and provision of moist environment for normal wound healing.

Keywords: Tissue adhesives, wound closures, fibrin sealants, cyanoacrylate adhesives

9.1 Introduction

Wound is externally induced injury to the skin or any tissue inside the body that results in breaks and damage to the body organs, cells, blood vessels, extracellular matrices, and membranes. Skin being outer covering acts as a shield to prevent body from injurious effects of environment. Successful wound healing process occurs in three overlapping stages inflammation and hemostasis, cell proliferation, and tissue remodeling [1], failure at any

**Corresponding author*: sajidakash@gcuf.edu.pk

Inamuddin, Rajender Boddula, Mohd Imran Ahamed and Abdullah M. Asiri (eds.) Green Adhesives: Preparation, Properties and Applications, (181–204) © 2020 Scrivener Publishing LLC

phase of wound healing process leads to serious complications. Wound healing and restoration of tissue functions is mandatory to avoid infection and bring the body to its original healthy condition [2]. Currently, research and interventions in wound closing devices lead to the development of wound healing adhesives to accelerate wound closure, and healing by acting as skin substitute, induce tissue growth, and reduce scar formation [3]. Wound scaffolds, dry gauzes, and bandages are commonly used [4] for wound closing but have demerits such as re-injury of tissues after their removal. Antibiotics containing wound dressings have chances of causing bacterial resistance and increase wound infections. To avoid such complications, nowadays, wounds are closed with staples, sutures, adhesive tapes, or wound adhesives according to their need as summarized in Table 9.1 [5, 6]. As low hydration requires more healing time, thus need of wound healing adhesives and their use in internal medicines with low toxicity profile, easy handling, cost effectiveness, prevention of fluid leakage from the sutures or incisions, provide moist healing (in case of hydrogels), and prevent undue tissue damage during removal [7] is of great interest for surgeons in clinical practices. Currently, scientists are searching alternative methods of wound closures without side effects for diabetic foot ulcer [8], endoscopic totally extraperitoneal (TEP) inguinal hernia repair [9], and incisions of hernia [10], surgical incisions [11], in ocular conditions like corneal incisions [12, 13], in retina [14], in children for facial laceration in emergency department [5], for oral wounds during dental surgeries [15], for gastrointestinal tract wound healing [16], for skin grafts in burns [17], as tissue adhesives in endoscopic bleeding [18], for total knee arthroplasty [19], and cardiac surgeries [20]. In literature, wound healing adhesives based upon their chemical nature such as albumin/glutaraldehyde sealant [20], gelatin-based adhesives [14], xylose-based semisynthetic polyurethane tissue adhesives [21], starch-based scaffolds [22], Poly(ethylene glycol) (PEG) dextran hydrogels [13, 23], PEG-based hydrogels [24], injectable self-healing hydrogels [3], polysaccharides-based wound adhesives [12], cyanoacrylates [5, 15, 25, 26], and fibrin sealants [17, 19] have been reported.

9.2 Wound

Wound is externally induced injury to the skin or any tissue inside the body that results in breaks and damage to the body organs, cells, blood vessels, extracellular matrices, and membranes. Wounds are classified as abrasion, avulsion, incision, laceration, and punctures (Figure 9.1).

Table 9.1 Summary of wound closures [6, 26].

	Advantages	Disadvantages	Indications	Contraindications
Sutures	Wound closure in less time and most meticulous approximation with low dehiscence rate is achieved. There is no need of microbe's removal.	May leave suture marks Require anesthesia Time-consuming method Induce infection and inflammation painful Risk of needle stick Cause nerve damage Granuloma and tissue scar formation Technical skills required require removal and cause re-injury	Animal or human bites Highly contaminated wounds Superficial wounds over joints Crushed or infected wounds Wounds at muco-cutaneous borders High tension areas wounds High moisture areas wounds 12 hour old Wounds Wounds at perineum, buttocks Laceration Incision	Infected and highly contaminated wounds
Staples	Fast Simple, easy Lower infection rates than sutures.	Painful Less meticulous and imprecise wound approximation Require removal Require use of anesthetics	Scalp wounds Large sized straight incisions and lacerations	Inadequate hemostasis Complex irregular wounds

(Continued)

Table 9.1 Summary of wound closures [6, 26]. (*Continued*)

	Advantages	Disadvantages	Indications	Contraindications
Tapes	Fast, Inexpensive, Simple, No risk of needle stick Easy method Low infection rates Minimal reactivity Comfortable No need for removal	Highest risk of dehiscence Low tensile strength Lowest tensile strength Not suitable on moist damp skin and hairy wound. No moisture resistant	Wound support after suture/ staple removal Linear incisions/wounds and lacerations with less tension and Fragile skin Under splints/casts	High tension Uncooperative Patients Hear bearing area Proximity to moist areas Circumferential use around digits
Tissue Adhesives	Fast Simple Virtually painless No needlestick required Dressing sloughs off Low cost Water proof dressing No anesthesia required	Avoid contact with eyes Heat discomfort Avoid contact with eyes Not for use on mucous membranes May bond with child Not proven on areas of tension	Linear low tension wounds and lacerations Fragile skin wounds Lower tensile strength Occlusive Dressing Fragile skin wounds	Inadequate hemostasis Incision Wounds at muco-cutaneous borders Wounds at perineum, buttocks High moisture areas wounds 12 hour old Wounds Hair bearing area Proximity to moist areas Inadequate hemostasis High tension

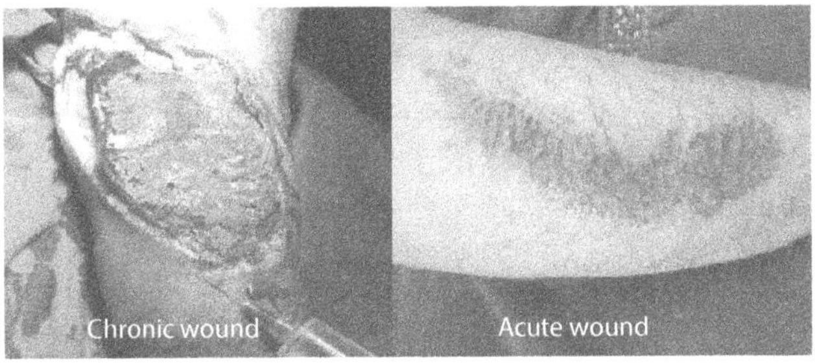

Figure 9.1 Acute and choric wounds.

9.2.1 Types of Wounds

9.2.1.1 Acute Wounds

Acute wounds are referred to those wounds that heal timely such as burns, traumatic injuries, and surgical incisions. Most acute wounds heal by natural healing mechanisms after proper wound closure.

9.2.1.2 Chronic Wounds

Chronic wounds are developed due to an abnormality in orderly and timely healing of wounds to redeem structural and functional integrity of tissues. Chronic wounds are often infected and exhibit a persistent anomalous inflammatory profile. Any skin lesion has the potential of becoming chronic, if not treated timely and properly. These wounds are named on the basis of their underlying cause. Common examples of chronic wounds are diabetic foot ulcers, pressure ulcers, and venous leg ulcers.

9.3 Structure and Function of the Skin

To understand the nature, treatment, and healing of external wounds, it is necessary to know about the structure and function of the skin. As an outer covering skin protect the body from physical damages, invading pathogens, fluid loss, hurtful substances in the environment (e.g., dust and microbial flora), and maintain the balance of body homeostasis. The skin is frequently damaged by external stress factors but it has the ability to restore itself without resulting in septicemia. Epidermis and the dermis are two layers of skin. The epidermis comprises of melanocytes, dendritic cells, keratinocytes, sensory axons, Langerhans cells, and basement membrane. The dermal layer contains skin appendages, fibroblasts, mast cells, immune cells, antigen containing dermal cells, and extra cellular matrix complex. Extra cellular matrix complex regulates functions of cytokine and growth factors and provides framework for intercellular connections, and allow cellular movement. Sensory autonomic nerve fibers form intercellular tight junctions with keratinocytes for transmission of pressure, pain, temperature, and itch sensations. Tonic adrenergic vasodilator and vasoconstrictor nerves control the arterial-venous thermoregulatory shunt circulation from which capillary network originates. Lymphatic vessels network is found under the epidermis. Pre-collector vessels is located in deeper layers of the dermis and lymph collecting vessels in the subcutaneous fat layer [1].

9.4 Mechanism of Wound Healing

Wound healing and restoration of tissue functions is a necessary process to bring the body to its original healthy condition. Wound healing process occurs in three sequential overlapping steps, inflammatory phase and hemostasis, proliferative phase, and tissue remodeling. For immediate stoppage of bleeding primary sensory, neurons sense the occurrence signs of injuries and send signals to the brain during inflammatory phase. Elimination of pathogens and wound cleaning by peripheral mononuclear blood cells is the preliminary target during inflammatory phase. After inflammatory phase and hemostasis proliferative phase starts, wounds repair and begin tissue remodeling. During this phase of wound repair, fibroplasia, angiogenesis, re-epithelialization, and peripheral nerve repair

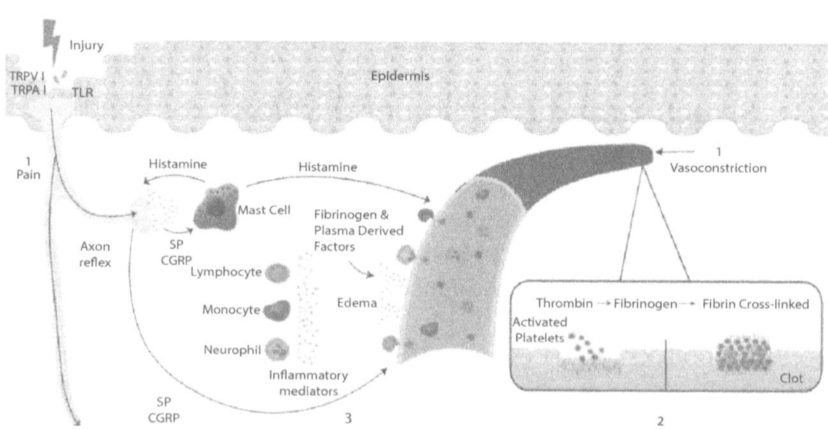

Figure 9.2 Mechanism of wound healing in skin adopted from Luis Cañedo-Dorantes and Mara Cañedo-Ayala [1].

occur. Completion of tissue remodeling and restoration of tissue integrity [1] are the main objectives of the final phase (Figure 9.2).

9.5 Wound Closing Techniques

After every surgical process, there is need of wound closing techniques to avoid infection and achieve proper wound approximation and wound healing. Since long time, wound closing is done by using sutures. Surgeons are looking for non-suture technique that maximizes tissue repair and minimizes infections by microbial flora. Every surgery leaves behind a scar of sutures that may have functional, cosmetic, or psychological consequences. Principle of wound closure should be ease in handling, precise wound approximation, and low infection rates. The ever-striving search for an alternative procedure to heal a wound with minimal scar, save time, and handle wound closing material easily lead to the discovery and development of synthetic sutures, staples, skin closure strips, clips, adhesive tapes, and tissue adhesives. Every patient wants a method that heals wound with better cosmetic appearance and minimal scar formation. The final decision of using appropriate wound closing technique rests on the surgeon. The surgeon must evaluate each case individually and choose wound closing technique which will maximize the opportunity of healing and minimize the likelihood of infection. Use of wound closing techniques, sutures, staples, tapes, and tissue adhesives depends upon the choice of surgeons as depicted in Table 9.1.

9.6 Wound Healing Adhesives

Wounds need to be closed to prevent infection and ensure proper healing and prevent infection [27]. Wound healing adhesives are materials that can stick to a wound tissue (e.g., connective, neural, epithelial, or muscular). These adhesives bind with injured tissues to stop bleeding, prevent fluid or gas leakages, and allow proper healing. Tissue adhesives were invented in 1949 and tried clinically for the first time in 1959. Since then, a number of wound healing adhesives have been developed by natural, derived, semisynthetic, or synthetic processes. This technique is used to close wound without using local anesthesia and allow normal wound healing by prohibiting entry of foreign material [28]. "Plaster of Paris" has been used around for centuries as tissue and bone adhesive. Wound closure and healing with adhesives remained a method of clinical interest that

Table 9.2 Role of wound healing adhesives.

Types of adhesives	Major role	Mechanism of action	Marketed product
Fibrin	Act as Haemostatic agent	Mimic blood coagulation	Tisseel™, Hemaseel™
Polyethylene glycol-based wound adhesives	Act as Haemostatic agent and important role in wound sealing	Improve water solubility	Duraseal™, Coseal®
Cyanoacrylates wound adhesives	Wound closure		Dermabond Advanced™
Polyurethanes wound adhesives	Wound closure		TissuGlu®
Albumin-based wound adhesives	Wound closure	Crosslinking with tissue protein	BioGlue® Progel™
Gelatin-based wound adhesives	Haemostatic agent	Swells in presence of blood and forms blood clot	Surgiflo® FloSeal Matrix®
Collagen-based wound adhesives	Wound closures Haemostatic when used in combination with thrombin	Formation of platelet plug	Floseal® Proceed® Costasis®
Chitosan-based wound adhesives	Wound closure	Enhance blood coagulation product Form bonding with tissue collagen due to presence of amine group through electrostatic attractions.	

can replace the conventional wound closing techniques such as staples, taps, and sutures in the repair of wounds.

The tissue adhesives have been used in general surgery to heal wounds related to eyes, kidney, liver, abdomen, and vascular surgeries in both pediatric and adults. Tissue adhesives are superior to conventional wound closing techniques bearing properties of easy to use as no technical skill required, comparable long-term cosmetic effects, decreased repair time, excellent bacteriostatic property, and elimination of recall visits. Adhesives carry no risk of sharps injury and prohibit the chance of infection and promote healing in topical lacerations and tension free small incisions. Sajid *et al.* in 2009 reviewed and compared skin adhesives versus sutures for closure of laparoscopic port-site wounds. Their statistical data has shown no difference between these two wound closing methods in terms of wound dehiscence, infection, and patient satisfaction. Wound healing adhesives were easy to use for closing port-site wound closure as compared with sutures [29]. Selection of wound closing technique depends upon the nature of wounds. Adhesives are not suitable for following wounds:

- Contaminated wounds from human or animal bites
- Infected or crushed wounds
- Wounds which cross muco-cutaneous borders
- Wounds over 12-hour old
- High friction areas such as buttocks
- Approximation of skin edges in wounds of deep layers require sutures
- Skin wounds over high tension areas such as joints
- Wounds at high moisture areas that allow poor adhesion [30]

Marketed products and role of wound healing adhesives are summarized in Table 9.2.

9.7 Types of Wound Healing Adhesives Based Upon Site of Application

Based upon biocompatibility and adhesive strength, tissue adhesives can be classified into two categories, namely, external use adhesives and internal use adhesives.

9.7.1 External Use Wound Adhesives

External use tissue adhesives are preferred over adhesive tapes, sutures, and staples and preferably used for sealing topical or local surgical incisions on or close to body surfaces due to their adhesive strength. During skin grafting, surgeons usually prefer external used adhesives for adhesion of artificial skin. Due to issue of biocompatibility, externally used tissue adhesives are not employed internally unless until no proven biocompatibility established, especially in those conditions that require direct contact of adhesives with the organs inside the body. Cyanoacrylate class of wound healing adhesives is employed externally and possesses the features of short gluing time, high adhesive strength, and ease in preparation. Due to production of heat during bonding process and formation of toxic degradation products such as formaldehyde and cyanoacetates, these adhesives are prohibited inside the body.

9.7.1.1 Steps for Applying External Wound Healing Adhesives on Skin [30]

1. Selection of wound to check suitability of adhesives according to the nature of wound
2. Clean wound with antiseptic solution
3. Connect skin edges by slight pulling on both ends
4. Apply adhesives to the wound in a way that it covers 5–10 mm either sides
5. Three coats should be applied
6. Allow 30–45 seconds for polymerization
7. Wound adhesive will peel of in 5–8 days
8. Do not speed up polymerization by fanning

9.7.2 Internal Use Wound Adhesives

Internal use tissue adhesives are used inside the body for fistulas occlusion, hemostasis during surgical processes, and adhesion sub-skin tissues, organs, and membranes. Use of internally used adhesives is comparatively less due to small displacement of organs and chances of toxic effects of cyanoacrylate glues. Biological compatibility is an important parameter for internally used tissue adhesives to be checked as per following criteria.

- Tissue adhesive and its degradation products should have no toxicity for human body.

- Adhesives should not be carcinogenic or inflammatory.
- Tissue adhesive should be irritation free and harmless to normal wound healing process.
- Material of tissue adhesives should be able to be degraded by hydrolysis and body enzymes.
- Metabolic products of tissue adhesives should be cleared from body by excretion.

9.8 Types of Wound Healing Adhesives Based Upon Chemistry

Schematic representation of wound healing adhesives based on chemistry is summarized in Figure 9.3.

9.8.1 Natural Wound Healing Adhesives

9.8.1.1 Fibrin Sealants/Fibrin-Based Tissue Adhesives

Fibrn sealant are biological glue widely used in dural closue and prevent from dural leakage, treatment of bronchopleural fistulas, prevention of air leakage during lungs procedures, cardiovascular surgery for hemostasis, to control bleeding from burns, avoid cerebrospinal fluid (CSF) leaks by providing water tight closure, and lacerations in spleen or liver. Use of fibrin-based adhesives in macular holes repairing, LASIK and scleral flaps, catract incisions, corneal ulcers, and transplants is of prime importance.

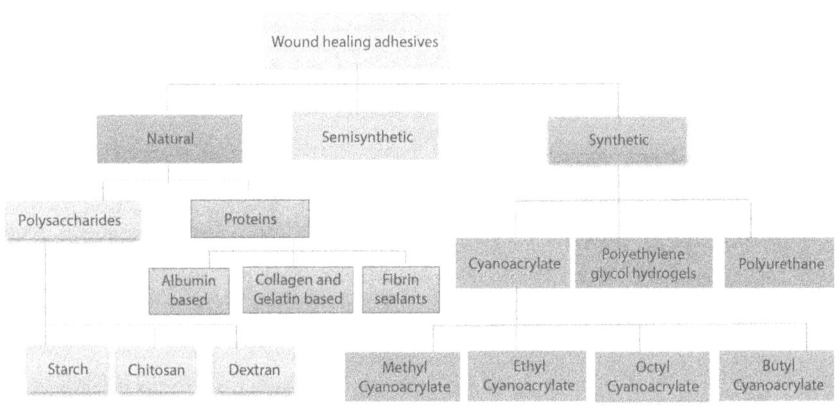

Figure 9.3 Classification of wound healing adhesives based upon chemistry.

Sealants are preferred over sutures and staples in surgeries where there is need of sealing air and gas leakages with minimal scarring. Esposito *et al.* in 2016 reviewed and summarized the literature to describe safety and efficacy of fibrin-based wound healing adhesives for sealing dura and prevent/treat cerebrospinal fluid leakages [31]. Due to their biologically safe nature, fibrin sealants have been used to decrease bleeding in knee replacements, radical neck dissections to treat lymphatic leaks. Physical and biological properties with better adhesive strength make tissue adhesives attractive for surgeons to use them in clinical practice for neurosurgery, periodontal, orthopaedic, endoscopic, ophthalmic, gastrointestinal, pneumothoracic, cardiovascular, plastic, and reconstructive surgeries. Due to moderate mechanical strength and good adhesive properties, fibrin sealants are mostly used with sutures. Hemostatic property and no heat generation enabled fibrin adhesives to be used in surgeries as shown by diagrammatic representation in Figure 9.4 adopted from [32].

Fibrinogen combination with thrombins started in 1944 as for using as an adhesive in skin grafts. Fibrin sealants are containing two components in double injector: fibrinogen with factor XIII and thrombin with Ca2+. Thrombin is helpful in clot formation by conversion of fibrinogen to fibrin monomer as it splits fibrinopeptide from α and β chains. Blood plasma fibrinogen plays an important role in blood clot formation. Thrombin activates fibrinogen and converts it into fibrin monomer when it is applied on

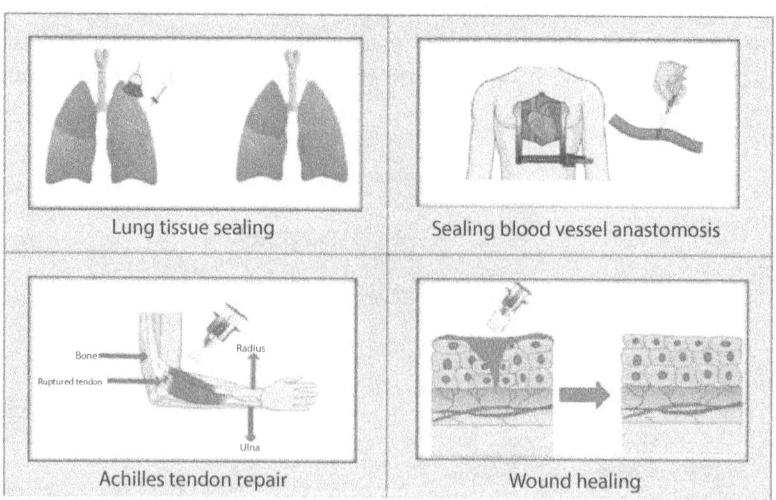

Figure 9.4 Use of surgical glues in surgical processes a diagrammatic representation adopted from [32].

wounds. While factor XIII, blood derived clotting proteins, enables fibrin monomers to crosslink and form insoluble clots. Formation of amide bonds between glutamine and lysine residues resulting in formation of an insoluble clot which is resistant to proteolytic cleavage. Calcium ion promotes these biological processes as shown in Figure 9.5 [11].

Thrombin and fibrin glues (or sealants) have not been approved for endoscopic procedures but are in use for blood clot. Al-Khazraji and Curry reviewed and highlighted clinical applications of tissue adhesives for acute variceal bleeding. They suggested that hemostasis can be achieved in variceal bleeding by direct injection of sclerosant and tissue adhesive into the varix during endoscopic ultrasound with low risk of hitches [18]. Mesh fixation to avoid recurrence of inguinal hernia after laparoscopic reparation is mostly done by staples, resulting in chronic inguinal pain. Choi *et al.* used fibrin glues for mesh fixation and observed its effect in reduction of acute or chronic postoperative pain in single-port laparoscopic totally extra peritoneal inguinal hernia repair patients. Daily living activities were resumed earlier and less anesthesia required in group of patients with mesh fixation by using fibrin glue as compared with mesh fixation by staples. No significant difference was observed between both groups in the incidence of chronic pain [33]. Fibrin sealants are familiarized to be used by some surgeons for the reduction of postoperative pancreatic fistula following major surgeries for cancer or inflammation of the pancreas. As the use of fibrin sealants following pancreatic surgery is controversial, Gong *et al.* used fibrin sealants in pancreatic surgery to assess the effectiveness, safety, and possible adverse effects of fibrin sealants for the

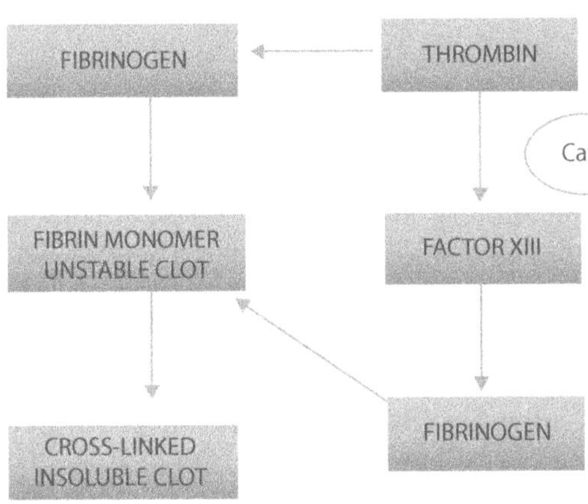

Figure 9.5 Diagrammatic representation of clot formation in fibrin glue adopted from [11].

prevention of postoperative pancreatic fistula. They reported little or no effect on pancreatic fistula after distal pancreatectomy [34]. Berney and Descallar performed an experiment for over a 10-year period and confirmed advantages and safety of fibrin glues as a substituent of staples. In their study, 703 patients underwent 1,000 elective totally extraperitoneal inguinal hernia repairs and fibrin glue mesh fixation was found extremely safe and reliable method, with a very high patient's satisfaction with limited risk of developing chronic pain [9]. Hovanesian suggested use of fibrin tissue adhesives in routine cataract surgeries to create a watertight seal with 7 days of adhesive strength more than enough for cataract surgery [35]. There was risk of hepatitis infection in fibrin treated patients but purification and sterilization techniques has expressively decreased the risk. The fibrin glue components are subjected to dry heat treatment, pasteurization, solvent-detergent cleansing, two step vapor heat treatment, precipitation, pH treatment, nanofiltration, and some chromatographic steps or combination of these processes for virus screening, inactivation or reduction, and make it useful in medical treatments [36].

9.8.1.2 Albumin-Based Adhesives

Albumin, globular protein naturally found in blood, used as hemostat in vascular and cardiac surgeries in combination with adhesion compound like glutaraldehyde. In combination form glutaraldehyde and bovine albumin adhesives are available in market in double chambered syringe approved is in the United States for assisting in the repair of aortic dissection. Glutaraldehyde covalently binds with proteins of wound and albumin molecules to form mechanical seal without affecting normal blood coagulation process. Upon adherence, Bioglu form interlocking within wound spaces [20]. Taguchi *et al.* in 2009 in their experiment combined organic acid-based crosslinkers disuccinimidyl tartarate, disuccinimidyl malate, and trisuccinimidyl citrate with human serum albumin to produce biodegradable, biocompatibility, and solid-liquid-type tissue adhesives with bonding time 5, 10, and 15min, respectively. It was found that bonding strength was dependent on the concentration and nature of crosslinker, as tissue adhesive having trisuccinimidyl citrate had shown 3.2-fold greater bonding strength than aldehyde-based tissue adhesives and 6-fold greater than fibrin-based adhesive [37]. Zhu *et al.* in 2017 prepared DOPA-albumin tissue adhesive for internal medical applications in which bovine serum albumin was acting as backbone, citric acid, and dopamine as additional components. DOPA albumin-based adhesive had 10-fold greater adhesion strength then commercially accessible fibrin-based glues in wet environment and fine *in vivo* biocompatibility, better *in vitro* gelation kinetics, swelling trends, degradation rates, and cytocompatibility [38].

In another study, Taguchi *et al.* combined human serum albumin with crosslinker citric acid derivative and prepared wound adhesives having bonding strength slightly low as compared to cyanoacrylates and 9 times greater than fibrin glues. Citric acid derived albumin adhesives has shown excellent wound closing property and biocompatibility similar to that of cyanoacrylate-based adhesives when used on mouse skin [39]. Nowadays, formaldehyde-based adhesives have not been approved by the FDA as formaldehyde is excluded from the Bioglu due to toxic properties.

9.8.1.3 Collagen and Gelatin-Based Wound Healing Adhesives

Collagen and gelatin-based adhesives are used in ocular applications [14] due to their excellent biocompatibility, ease of processing, effective crosslinking properties, and low cost. Gelatin scaffolds with other polymers have widely used as bio-artificial grafts, bio-adhesives, and cell-sheet carriers during last decade [40]. *In situ* forming sprayable and injectable gelatin-based hydrogels have been emerged as an substitute to conventional wound tissue adhesives [41]. The collagen-based adhesives may have beneficial effects in medical treatments as a sealant film, for bonding soft tissues, tendon wraps for preventing adhesion following surgery, and wound closures [42]. Collagen-based adhesives Floseal (Sulzer Spine-tech, Anaheim, California) are used for vascular surgeries and Proceed (Fusion Medical Technologies, Mountain View, California) used for prevention and treatment of CSF leakage, Costasis (Cohesion Technologies, Inc.), used to control bleeding in of hepatic, orthopedic, and cardiovascular surgeries. All three adhesives are made by combining bovine and thrombin bovine collagen.

9.8.1.4 Starch

Starch is mostly used in the formation of scaffolds, tissue engineering, bone replacement implants, and different kinds of wound dressings. Salgado *et al.* evaluated the *in vivo* response of starch-based scaffolds when implanted in rats and concluded these scaffolds have remarkable effects in very early form of bone formation [43]. Different starch-based wound dressings are prepared by its modifications and reviewed in literature [44]. Marques *et al.* in 2002 reported the cytocompatibility of starch-based polymers as potential biomaterials. The biocompatibility of different blends of corn-starch, starch/cellulose acetate, and starch/ethylene vinyl alcohol were assessed. Both types of starch-based polymers exhibit a cytocompatibility that might allow for their use as biomaterials [45].

9.8.1.5 Chitosan

Chitosan form bonding with tissue collagen due to presence of amine group through electrostatic attractions. Chitosan-based wound healing adhesives besides increasing blood coagulation process can also releases transforming and platelet-derived growth factors which play an essential role in the wound healing process [46]. Polysaccharide-based biocompatible and biodegradable polymer chitosan, composed of N-Acetyl glucosamine, have emerged as wound adhesives without the production of harmful byproducts and undergoes enzymatic degradation in biological environment. These chitosan/pluronic-based hydrogels have proved its usefulness in tissue engineering by injectable drug delivery and act as antibleeding ingredients [47]. Chitosan has not been commercialized and marketed on larger extent due to its poor water solubility but it can lead to the discovery of non-toxic biodegradable and biocompatible wound healing adhesives [48]. Chitosan-PEG-tyramine (CPT) hydrogels were prepared by using hydrogen peroxide and horseradish peroxidase to explore their tissue adhesives characteristics. PEG was grafted on chitosan backbone to crosslink in three-dimensional network and enhance its solubility. CPT gels had shown best results in healing when compared with fibrin glue, cyanoacrylates, and suture. They suggested that the formation of *in situ* remediable chitosan hydrogels can lead to the invention of very interesting and promising wound healing adhesives [49]. Hydrophobic nature of chitosan makes it able to form cluster of blood cells by connecting them in wounds and stop bleeding and can be used as sprayable foam-based sealants. The hemostasis with chitosan-based adhesives can be achieved within minutes after its application on wounds and blood loss can be 90% reduced [50].

9.8.1.6 Dextran

Bhatia *et al.* in 2007 used a combination of dextran aldehyde with star polyethylene amines for sealing corneal incision in rabbit eyes. These wound healing adhesives were found efficacious for sealing corneal incisions without cytotoxic effect on corneal endothelial cells [12]. Matsumura *et al.* prepared oxidized dextran and Poly-L-lysine hydrogel. They have reported well-controlled degradation rate, low toxicity, and potential clinical applications of above mentioned bioadhesives [51]. Wang *et al.* in 2012 prepared photocrosslinkable gels from bio-degradable and biocompatible polymers, gelatin, and oxidized urethane dextran. They suggested that gelatin and dextran (double crosslinker)-based gels can be used as a novel wound healing adhesives for better cell attachment in such a way that gelatin component can decrease the swelling ratio and increase the adhesive strength [52]. Chenault *et al.* reported tissue adhesive properties of combination of 8-armed star poly(ethylene glycol) with (MW 10,000), bearing primary amine groups on each arm and 50% oxidized dextran aldehyde (MW 10,000). They found that dextran aldehyde in combination with 8-armed star poly(ethylene glycol) and tissue adhesives cures spontaneously without cytotoxicity and have strong bond forming capability that only 1–2 microliter is enough for clear corneal incision sealing within first 5 days [13]. Artzi *et al.* in 2009 reported biocompatible PEG/dextran-based adhesives with amine groups are superior to cyanoacrylates wound adhesives having comparatively better adhesive properties than fibrin-based wound adhesives [23].

9.8.2 Synthetic Wound Healing Adhesives

9.8.2.1 Cyanoacrylate

Cyanoacrylates commonly known as glue or superglue were used for wound treatment despite of disapproval from FDA until 1998. They tend to get polymerized when make contact with skin by forming mighty bio films due to strong bonding with wound and endergonic reactions as well. At last, their use came to end for their weak tensile strength and brittleness. Now, their use for the treatment of oral wounds due to dental surgery has also been demonstrated [15]. This class of wound healing adhesives is employed externally and possesses the features of short gluing time, high adhesive strength, and ease in preparation. Toxic degradation products such as formaldehyde and cyanoacetates and heat production during bonding lemmatized their use inside the body. To minimize the toxic effects and improve acceptability derivatives of cyanoacrylates has been developed. Methyl 2-cyanoacrylate, Ethyl-2-cyanoacrylate, 2-butyl cyanoacrylate, and 2-octyl cyanoacrylate are important derivatives of this class of wound adhesives (Figure 9.6). Singh *et al.* in 2019 evaluated effectiveness of n-butyl cyanoacrylates wound healing adhesives in maxillofacial and oral laceration after surgical incisions [53] and reported that adhesives are cost effective and easy to apply. Singer *et al.* in 2008 reviewed topical applications of cyanoacrylates [26]. Soni *et al.* in 2013 studied the effectiveness of 2-octyl cyanoacrylates tissue adhesive in maxillofacial lacerations and

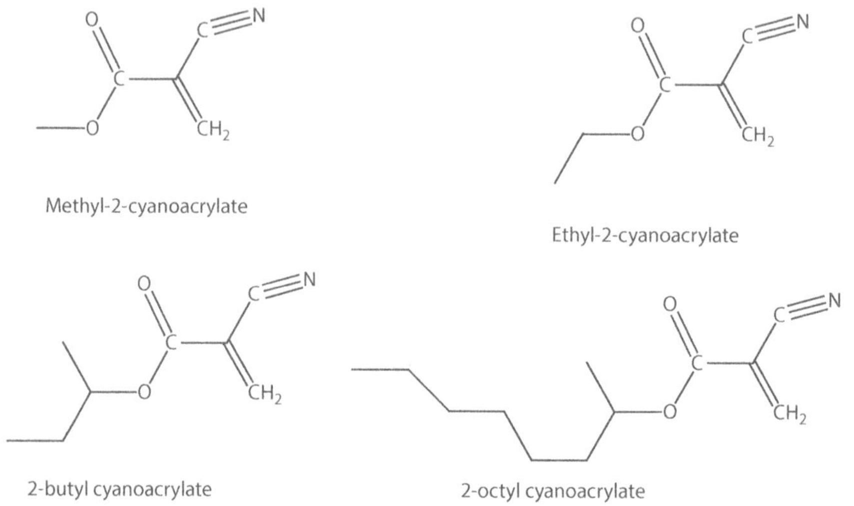

Figure 9.6 Structures of cyanoacrylates.

reported that these adhesives can be used for fast, effective, reliable skin closure in maxillofacial incisions [54].

9.8.2.2 *Poly Ethylene Glycol-Based Wound Adhesives (PEG)*

Due to large weight and hydrophilic properties of PEG-based wound adhesives especially when PEG form back bone with macromers, these adhesives possess high swelling characteristics and exert pressure on surrounding tissues. Strehin *et al.* in 2009 synthesized PEG-based wound adhesives to apply on corneal incisions in combination chondroitin sulfate and characterized *in vivo* and *in vitro* biological and physical properties to restore intra ocular pressure. These wound adhesives were biocompatible with corneal tissues. These adhesives had no adverse effects in live swine model [55]. Focalseal-L is an FDA approved marketed adhesive belonging to this group, its activation and polymerization is done by light source and give better adhesion. Degradation time of focalseal-L is almost 3 months makes it suitable for use in thoracic surgery to decrease air leaks and but these adhesives are not suitable in hemorrhages.

9.8.2.3 *Hydrogels*

Hydrogels are mostly used as extra cellular matrix to close LASKI flaps, corneal incisions, and perforations. Grinstaff in 2007 prepared *in situ* polymerizing hydrogel-based adhesives with dendrimers by peptide ligation reaction that has corneal wound adhesive property after photo-crosslinking reaction. There are chances of retinal damage due to light during photocrosslinking [56]. Self-gelling hydrogels having crosslinking property by peptide ligation do not require light. Biocompatible, highly branched dendritic macromolecules (with multiple ends) can be used to form hydrogels with wounds. Xu *et al.* prepared polysacharides-based hydrogels that crosslinked through reversible covalent/noncovalent reactions having stress relaxation and self-healing properties. They changed their properties by incorporating electroconductive polymers exhibiting reliable wound adhesive strength comparable with fibrin glue. These cytocompatibil adhesives were soft and had characteristics of increasing myogenic differentiation in myoblast cells. These hydrogel adhesives when comprising of heparin could be used as injectable drug delivery in cardiovascular diseases and wound closing [57]. Hydrogels exhibit tissue adhesive properties by interactions of incorporating functional groups with the surrounding tissues and formation of interpenetrating networks to close the injured sites as shown in Figure 9.7. Designing of hydrogels depends upon the nature and location of the wound and its neighboring tissues and fluids. Characterization is done by measuring mechanical, chemical, physical, and adhesive properties. Hydrogels have the properties to adhere in moist condition without affecting normal healing mechanism that make it acceptable in the field of wound healing adhesives.

The hydrogel components must be safe and compatible with the biological system to allow proper wound healing. Thus, the utilization of biocompatible and biodegradable polymer structures is of utmost importance in the design of hydrogel adhesives. Hydrogel must adhere to wound sites and completely seal it to prevent fluid leakages and infection by bacterial penetration. There should be no heat generation. Hydrogels should have properties of biodegradability and safe degradation products and can be eliminated from body after wound healing. Potential toxic effects can be minimized by formulating hydrogels at physiological pH having buffer capacity.

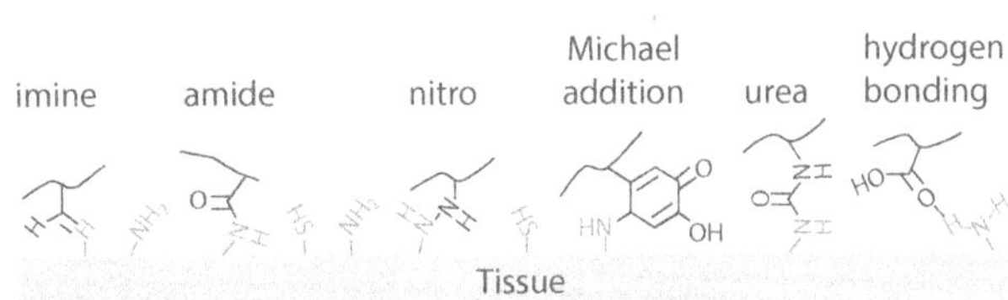

Figure 9.7 Diagramatic representation of chemical/physical linkages between tissues and hydrogels adopted from [60].

9.8.2.4 Polyurethane

During 1960s, biostable polyurethanes (PURs) have emerged to be incorporated in biomedical devices. Currently, biodegradable polyurethanes have been investigated for their use in regenerative medicines. Their degradation to non-toxic decomposition products, *in vivo* biostability and ability to promote growth of tissues and cells *in vivo* and *in vitro* is of keen interest. Biological, mechanical, and physicochemical properties enable the biodegradable polyurethanes to be used as scaffolds for regeneration of tissue in future [58]. Wound healing process was improved in animal tissues by applying tough but elastic coating of a mixture of polyether polyurethane resin pellets in tetrahydrofuran [59]. Balchioglu *et al.* in 2016 performed an experiment and synthesized semisynthetic xylose-based polyurethane (NPU-PEG-X) bioadhesives to study protein adsorption, adhesion properties, *in vitro/in vivo* biocompatibility and *in vitro* biodegradation, properties of these newly formed bio adhesives. They reported that these bio adhesives could play a significant role as supporting materials in medical field to prevent fluid leakage after wound closure and abdominal surgeries [21].

9.9 Summary

Review of wound healing adhesives may lead to the discovery of biodegradable, biocompatible, non-toxic wound closing techniques to be applied both inside and outside the body. Diabetes, a metabolic disorder, may increase the complications of wound healing process leads to the formation of chronic wounds such as diabetic foot ulcer, pressure ulcers, and venous leg ulcers. There is great need of wound healing adhesives that may heal and seal wounds without technical skill, tissue re-injury, and, in less time, and cost.

References

1. Canedo-Dorantes, L. and Canedo-Ayala, M., Skin Acute Wound Healing: A Comprehensive Review. *Int. J. Inflam.*, 2019, 3706315, 2019.
2. Eming, S.A., Martin, P., Tomic-Canic, M., Wound repair and regeneration: Mechanisms, signaling, and translation. *Sci. Transl. Med.*, 6, 265, 265sr6, 2014.
3. Chen, H. *et al.*, An injectable self-healing hydrogel with adhesive and antibacterial properties effectively promotes wound healing. *Carbohydr. Polym.*, 201, 522–531, 2018.

4. Boyadzhieva, S. *et al.*, A Self-Adhesive Elastomeric Wound Scaffold for Sensitive Adhesion to Tissue. *Polymers (Basel)*, 11, 6, E942, 2019.
5. Singer, A.J. *et al.*, Patterns of use of topical skin adhesives in the emergency department. *Acad. Emerg. Med.*, 17, 6, 670–2, 2010.
6. Mattick, A., Use of tissue adhesives in the management of paediatric lacerations. *Emerg. Med. J.*, 19, 5, 382–5, 2002.
7. Demina, N.B., Chernova, L.V., Kozlova, Z.M., Application of adhesive compositions in surgery (in Russian only)]. *Khirurgiia (Mosk)*, 3, 3, 129–134, 2019.
8. Andrews, K.L., Houdek, M.T., Kiemele, L.J., Wound management of chronic diabetic foot ulcers: From the basics to regenerative medicine. *Prosthet. Orthot. Int.*, 39, 1, 29–39, 2015.
9. Berney, C.R. and Descallar, J., Review of 1000 fibrin glue mesh fixation during endoscopic totally extraperitoneal (TEP) inguinal hernia repair. *Surg. Endosc.*, 30, 10, 4544–52, 2016.
10. Brown, J.K. *et al.*, A prospective, randomized comparison of skin adhesive and subcuticular suture for closure of pediatric hernia incisions: Cost and cosmetic considerations. *J. Pediatr. Surg.*, 44, 7, 1418–22, 2009.
11. Bhagat, V. and Becker, M.L., Degradable Adhesives for Surgery and Tissue Engineering. *Biomacromolecules*, 18, 10, 3009–3039, 2017.
12. Bhatia, S.K. *et al.*, Polysaccharide-based tissue adhesives for sealing corneal incisions. *Curr. Eye Res.*, 32, 12, 1045–1050, 2007.
13. Chenault, H.K. *et al.*, Sealing and healing of clear corneal incisions with an improved dextran aldehyde-PEG amine tissue adhesive. *Curr. Eye Res.*, 36, 11, 997–1004, 2011.
14. Chen, T. *et al.*, Gelatin-based biomimetic tissue adhesive. Potential for retinal reattachment. *J. Biomed. Mater. Res. Part B: Appl. Biomater.: An Official Journal of The Society for Biomaterials, The Japanese Society for Biomaterials, and The Australian Society for Biomaterials and the Korean Society for Biomaterials*, 77, 2, 416–422, 2006.
15. Borie, E. *et al.*, Oral Applications of Cyanoacrylate Adhesives: A Literature Review. *Biomed. Res. Int.*, 2019, 8217602, 2019.
16. Nishiguchi, A., Kurihara, Y., Taguchi, T., Underwater-adhesive microparticle dressing composed of hydrophobically-modified Alaska pollock gelatin for gastrointestinal tract wound healing. *Acta Biomater.*, 99, 387–396, 2019.
17. Miller, R. *et al.*, Systematic review of fibrin glue in burn wound reconstruction. *Br. J. Surg.*, 106, 3, 165–173, 2019.
18. Al-Khazraji, A. and Curry, M.P., The current knowledge about the therapeutic use of endoscopic sclerotherapy and endoscopic tissue adhesives in variceal bleeding. *Expert Rev. Gastroenterol. Hepatol.*, 13, 9, 893–897, 2019.
19. Wang, H. *et al.*, Is fibrin sealant effective and safe in total knee arthroplasty? A meta-analysis of randomized trials. *J. Orthop. Surg. Res.*, 9, 36, 2014.
20. Chao, H.H. and Torchiana, D.F., BioGlue: Albumin/glutaraldehyde sealant in cardiac surgery. *J. Card. Surg.*, 18, 6, 500–503, 2003.

21. Balcioglu, S. *et al.*, Design of xylose-based semisynthetic polyurethane tissue adhesives with enhanced bioactivity properties. *ACS Appl. Mater. Interfaces*, 8, 7, 4456–4466, 2016.
22. Armour, W.B. and Kumar, R.-C.D., Water resistant adhesives based on acetoacetate esters of starch, 1968, Google Patents.
23. Artzi, N. *et al.*, Characterization of star adhesive sealants based on PEG/dextran hydrogels. *Macromol. Biosci.*, 9, 8, 754–765, 2009.
24. Chen, K. *et al.*, Entanglement-Driven Adhesion, Self-Healing, and High Stretchability of Double-Network PEG-Based Hydrogels. *ACS Appl. Mater. Interfaces*, 11, 40, 36458–36468, 2019.
25. Blondeel, P.N. *et al.*, Closure of long surgical incisions with a new formulation of 2-octylcyanoacrylate tissue adhesive versus commercially available methods. *Am. J. Surg.*, 188, 3, 307–313, 2004.
26. Singer, A.J., Quinn, J.V., Hollander, J.E., The cyanoacrylate topical skin adhesives. *Am. J. Emerg. Med.*, 26, 4, 490–6, 2008.
27. Farion, K.J. *et al.*, Tissue adhesives for traumatic lacerations in children and adults. *Cochrane Database Syst. Rev.*, 3, 3, CD003326, 2002.
28. Lauto, A. *et al.*, *In vitro* and *in vivo* tissue repair with laser-activated chitosan adhesive. *Lasers Surg. Med.: The Official Journal of the American Society for Laser Medicine and Surgery*, 39, 1, 19–27, 2007.
29. Sajid, M.S. *et al.*, Meta-analysis of skin adhesives versus sutures in closure of laparoscopic port-site wounds. *Surg. Endosc.*, 23, 6, 1191–1197, 2009.
30. Marshall, G., Skin glues for wound closure. *Aust. Prescr.*, 36, 2, 49–51, 2013.
31. Esposito, F. *et al.*, Fibrin Sealants in Dura Sealing: A Systematic Literature Review. *PLoS One*, 11, 4, e0151533, 2016.
32. Jain, R. and Wairkar, S., Recent developments and clinical applications of surgical glues: An overview. *Int. J. Biol. Macromol.*, 137, 95–106, 2019.
33. Choi, B.J., Jeong, W.J., Lee, S.C., Fibrin glue versus staple mesh fixation in single-port laparoscopic totally extraperitoneal inguinal hernia repair: A propensity score-matched analysis. *Int. J. Surg.*, 32–37, 2018.
34. Gong, J. *et al.*, Fibrin sealants for the prevention of postoperative pancreatic fistula following pancreatic surgery. *Cochrane Database Syst. Rev.*, 6, Cd009621, 2018.
35. Hovanesian, J.A. and Karageozian, V.H., Watertight cataract incision closure using fibrin tissue adhesive. *J. Cataract Refract. Surg.*, 33, 8, 1461–3, 2007.
36. Sierra, D.H. *et al.*, A method to determine shear adhesive strength of fibrin sealants. *J. Appl. Biomater.*, 3, 2, 147–51, 1992.
37. Taguchi, T. *et al.*, Biodegradable adhesives composed of human serum albumin and organic acid-based crosslinkers with active ester groups. *J. Bioact. Compat. Polym.*, 24, 6, 546–559, 2009.
38. Zhu, W. *et al.*, A novel DOPA-albumin based tissue adhesive for internal medical applications. *Biomaterials*, 99–115, 2017.

39. Taguchi, T. *et al.*, Development of a novel glue consisting of naturally-derived biomolecules: Citric acid and human serum albumin. *J. Nanosci. Nanotechnol.*, 7, 3, 742–747, 2007.
40. Rose, J.B. *et al.*, Gelatin-based materials in ocular tissue engineering. *Materials*, 7, 4, 3106–3135, 2014.
41. Lee, Y. *et al.*, *In situ* forming gelatin-based tissue adhesives and their phenolic content-driven properties. *J. Mater. Chem. B*, 1, 18, 2407–2414, 2013.
42. Kelman, C.D. and DeVore, D.P., Method for sealing tissues with collagen-based sealants, 1999, Google Patents.
43. Salgado, A. *et al.*, *In vivo* response to starch-based scaffolds designed for bone tissue engineering applications. *J. Biomed. Mater. Res. Part A*, 80, 4, 983–989, 2007.
44. Torres, F.G., Commeaux, S., Troncoso, O.P., Starch-based biomaterials for wound-dressing applications. *Starch-Stärke*, 65, 7-8, 543–551, 2013.
45. Marques, A., Reis, R., Hunt, J., The biocompatibility of novel starch-based polymers and composites: *In vitro* studies. *Biomaterials*, 23, 6, 1471–1478, 2002.
46. Okamoto, Y. *et al.*, Effects of chitin and chitosan on blood coagulation. *Carbohydr. Polym.*, 53, 3, 337–342, 2003.
47. Ryu, J.H. *et al.*, Catechol-functionalized chitosan/pluronic hydrogels for tissue adhesives and hemostatic materials. *Biomacromolecules*, 12, 7, 2653–2659, 2011.
48. Ono, K. *et al.*, Photocrosslinkable chitosan as a biological adhesive. *J. Biomed. Mater. Res.: An Official Journal of The Society for Biomaterials and The Japanese Society for Biomaterials*, 49, 2, 289–295, 2000.
49. Lih, E. *et al.*, Rapidly curable chitosan–PEG hydrogels as tissue adhesives for hemostasis and wound healing. *Acta Biomater.*, 8, 9, 3261–3269, 2012.
50. Dowling, M.B. *et al.*, Sprayable foams based on an amphiphilic biopolymer for control of hemorrhage without compression. *ACS Biomater. Sci. Eng.*, 1, 6, 440–447, 2015.
51. Matsumura, K. *et al.*, Self-degradation of tissue adhesive based on oxidized dextran and poly-L-lysine. *Carbohydr. Polym.*, 113, 32–38, 2014.
52. Wang, T., Nie, J., Yang, D., Dextran and gelatin based photocrosslinkable tissue adhesive. *Carbohydr. Polym.*, 90, 4, 1428–1436, 2012.
53. Singh, P.K. *et al.*, To Evaluate the Efficacy and Effectiveness of N-butyl-2-cyanoacrylate glue (TRU SEAL) in Closure of Oral and Maxillofacial Laceration and Surgical Incisions. *J. Maxillofac. Oral Surg.*, 18, 1, 131–138, 2019.
54. Soni, A. *et al.*, Comparing cyanoacrylate tissue adhesive and conventional subcuticular skin sutures for maxillofacial incisions–a prospective randomized trial considering closure time, wound morbidity, and cosmetic outcome. *J. Oral Maxillofac. Surg.*, 71, 12, 2152.e1–8, 2013.

55. Strehin, I. *et al.*, Synthesis and characterization of a chondroitin sulfate-polyethylene glycol corneal adhesive. *J. Cataract Refract. Surg.*, 35, 3, 567–576, 2009.
56. Grinstaff, M.W., Designing hydrogel adhesives for corneal wound repair. *Biomaterials*, 28, 35, 5205–5214, 2007.
57. Xu, Y. *et al.*, Cytocompatible, Injectable, and Electroconductive Soft Adhesives with Hybrid Covalent/Noncovalent Dynamic Network. *Adv. Sci. (Weinh.)*, 6, 15, 1802077, 2019.
58. Guelcher, S.A., Biodegradable polyurethanes: Synthesis and applications in regenerative medicine. *Tissue Eng. Part B: Rev.*, 14, 1, 3–17, 2008.
59. Huprich, C.A., Method and composition for coating a wound with polyether polyurethane, 1993, Google Patents.
60. Ghobril, C. and Grinstaff, M., The chemistry and engineering of polymeric hydrogel adhesives for wound closure: A tutorial. *Chem. Soc. Rev.*, 44, 7, 1820–1835, 2015.

10

Green-Wood Flooring Adhesives

Mustafa Kucuktuvek

Department of Interior Architecture and Environmental Design, Faculty of Fine Arts and Architecture, Antalya Bilim University, Antalya, Turkey

Abstract

The protection of human health and nature is becoming increasingly important in interior architecture. Because of this reason, the non-toxic adhesives used for wood flooring became more important. Green adhesives used in wood flooring based on green building, eco-design, and sustainability principles are examined in this book chapter. Green buildings can provide a superior indoor environment. Adhesives used in wood flooring must have low or zero volatile organic compounds (VOCs) in order to improve indoor air quality in architecture. Today, most of the adhesives are prepared from petroleum-based polymers; however, these adhesives forms release formaldehyde and VOC emission. Increased environmental awareness led to the search for healthy adhesive alternatives with low emission values. Research centers have therefore sought to discover healthy adhesives by working on mostly bio-based adhesives. This book chapter discusses synthetic or renewable adhesive systems that can replace formaldehyde-based adhesives. In this book chapter, recent researches about wood flooring adhesive made in recent years are included.

Keywords: Wood flooring, green adhesives, green building, eco-design, interior architecture

10.1 Introduction

Healthy living space is possible with green buildings. Healthy interiors have a positive effect on living comfort and working performance. To improve indoor air quality, wood flooring adhesives with zero or low volatile compounds (VOC) must be used. Not only adhesives but also other

Email: mustafa.kucuktuvek@antalya.edu.tr; intercontinental@windowslive.com

Inamuddin, Rajender Boddula, Mohd Imran Ahamed and Abdullah M. Asiri (eds.) Green Adhesives: Preparation, Properties and Application, (205–226) © 2020 Scrivener Publishing LLC building materials and cleaning and maintenance products can cause toxic gas release [1]. The life cycle of the materials and building are important for ecological design. The ecological design includes a life cycle assessment (LCA). In case of excessive efforts, LCA is considered among the effective methods for determining environmental impacts. The United Nations environmental program also has views on environmental impacts [2]. Materials used indoors must have a vision throughout their lifetime. When designing, it must ensure that environmental impacts are identified as well as serious prevention of these impacts [3]. Ecological designs cannot be made with a focus on environmental impacts only. However, LCA focuses on the environmental impacts of the building system. With eco-design, multiple tools are handled. Eco-design is a lifecycle and incorporates economic performance, technical feasibility, and social practices and sustainability. The life-cycle assessment uses well-known tools to determine the environmental profile of the product [4].

Adhesives have an indispensable role in the forest products industry and in flooring. The development of adhesives will also help the development of the forest products industry [5, 6]. Different pieces of wood are used in furniture and flooring. Adhesives combine these parts to produce modern and functional products [7, 8]. For centuries, organic adhesives have been used in floor coverings. However, with the development of the industry, thermosetting adhesives have begun to be used because of their advantages [5].

Wood adhesives from polymers of petroleum derivatives have been widely used. However, the increased sensitivity to the environment and human health allowed the development of new alternative adhesives made from biological sources. New adhesives produced from biological sources must have high adhesion and bonding properties in order to compete with thermosetting adhesives. It should be resistant to water and heat and should be low cost [9]. Today, most phenolic resins and amino-based isocyanates are used as thermoset adhesives. Such adhesives are used in wood flooring as reactive and more economical adhesives due to their curing properties. Such adhesives have been widely used in the wood composite industry for decades [10]. The urea-formaldehyde adhesive is widely used in the production of medium-density fiberboard (MDF) and particleboards. This is mainly due to the high adhesion resistance and colorlessness of this adhesive [11].

Twenty million cubic meters of formaldehyde-based adhesive is used in the world in 2010. The threat of formaldehyde emissions has been known since the 1970s and was under constant pressure to reduce it. It was recently classified as a category 1B carcinogenic compound by Europe in 2016. Increased pressure due to the harmful effects of formaldehyde makes it necessary to find new alternatives [12]. Urea Formaldehyde adhesives are the most commonly used adhesive [13]. The reason for the use of urea-formaldehyde adhesive is its low cost.

Debates on emissions of wood-based compounds began in the 20th century as reviewed by Roffael *et al.* [14, 15]. Problems with formaldehyde emissions are also more common in interior spaces. These problems have been discussed by some authors according to Salthammer *et al.* [16].

The emissions within the polymer network vary depending on the type of formaldehyde bond [17]. In the 1970s, formaldehyde emissions were found to have carcinogenic effects on mice. In the late 1980s, ecology-oriented construction institutes began to limit emissions from formaldehyde [18].

Hemicellulose, lignin, and cellulose, one of the constituents of wood, naturally emit formaldehyde emissions [19] in addition to its extracts [20]. Efforts are underway to reduce the emissions of adhesives used in floor coverings in Japan and America, which are among the developed countries in the world. Different authorities carry out studies on this subject [21], risks on this issue have been discussed in different studies [22]. In a study, the "International Agency for Research on Cancer" (IARC) reported that formaldehyde was categorized as a category 1 carcinogenic substance [22].

Low formaldehyde emissions are defined in the "JIS A 5905 Japanese or JAS SE 9" standards [23–25].

Another way to reduce formaldehyde emissions may be to increase the number of proteins added to the resins. These kinds of resins are available in Germany [26–28]. The hydrolytically stable phenol bonds are formed by the addition of proteins to the thermoset glue. Therefore, lower formaldehyde emissions occur [18, 26]. Formaldehyde emission can be released in different ways from the wood flooring materials [29].

Formaldehyde will be banned in Europe for many health-threatening reasons, mainly because it is carcinogenic and allergic. Environmental awareness in the world is gradually increasing, and as a result, the production of adhesives that do not threaten human health is inevitable. Before the introduction of petroleum-based adhesives, biopolymers such as starch were used. The most important feature of synthetic adhesives is that they are water-resistant. Green wood-flooring adhesive can be produced from renewable sources, especially agricultural and forest resources. R & D studies on this subject have started [30].

Formaldehyde-free adhesives and adhesives with low formaldehyde release are available. Formaldehyde-free adhesives are characterized by the absence of formaldehyde incorporated into the network. The difference between phenolic resins and aminoplast adhesives is that they are hydrolytically stable, but urea-formaldehyde resins have much formaldehyde release compared to phenol-formaldehyde resins [31].

The main reason for the use of fossil-based adhesives in wood flooring is its high bond strength and water resistance. In addition, their low cost has played an important role in choosing fossil-based adhesives [32]. Protein-based adhesives should be economical because fossil-based adhesives are harmful to human health. In addition, adhesion resistance and water resistance should be improved, so extensive research is underway [33–35]. The main reason why soy proteins are shown as an alternative to fossil-based adhesives is that they are cost-effective, easy to process, and readily available. However, soy proteins are also a source of nutrients, although adhesives made from soy proteins have been proposed for use in wood flooring [33, 36]. Protein-based adhesives were added to phenolic adhesives and polyvinyl acetate (PVAc) glue to reduce formaldehyde emissions [34, 37]. A green route has been demonstrated to prepare a soy protein-based adhesive. Proteins and polysaccharides are first separated to form intertwined bonds. The vial is then cross-linked with sodium hexametaphosphate with polysaccharides, a cross-linker, and the final proteins are mixed together. This adhesive is a stable thermosetting adhesive that can be used in Wood flooring [38]. Lignin can be added to soy protein-based adhesives to increase the water-resistance [39]. Wood flooring adhesive water resistance can be increased by small particle size lignin. Therefore, on the bonding performance of the adhesive, protein lignin ratio, and lignin particle size have a great effect [40]. In a study before reductive amination lignin amine, fenton oxidation was reported to be prepared [41]. Cottonseed protein isolates and soy protein were mixed in different proportions to produce wood glue. However, increasing the fraction of soy protein had a negative effect on adhesion properties. Cellulose, starch, and xylan were added to the formulation to increase the adhesion strength to hot water. It is also known that the lignin amine enhances the binding performance and water-resistance of the soy protein [42]. Soybean crops cannot be produced all over the world due to climatic conditions. Therefore, wheat gluten, which is one of the by-products of wheat starch and bioethanol production, is an alternative wood adhesive [43–53].

Polysaccharides are an interesting polymeric wood binder alternative with binding properties [54]. A high molar mass means high viscosity, but a high molar mass polysaccharide provides high adhesion. The hydrogen bonds between in the polysaccharide and the substrate provide high bonding strength. Hydroxyl groups are not water-resistant. The resistance to water needs to be increased [30].

Among the polysaccharides, the polysaccharide has the potential to be the largest wood adhesive [34, 35, 55]. Starch is one of the natural adhesive sources and consists of amylopectin and amylose, and the ratio of amylopectin and amylose affects the adhesion properties of the starch-based adhesive. It is one of the most important features of being a renewable substance and being widely available and being a cheap raw material. Starch is an alternative to a good adhesive, but its resistance to water needs to be increased so that it is tried to increase its resistance to water with the help of polyvinyl acetate and polyvinyl alcohol [56, 57].

Naturally found only in Mucoraceae mushrooms, chitosan is also found in crustaceans. Shrimp and crab are the examples. Chitosan and chitin are thought to be synthesized in related organisms at a rate of 10 billion tons per year [58]. Important properties of chitosan: non-toxic, biodegradable, and reactive amino groups characterized by a high molar mass allowing for antimicrobial and chemical modification. Deacetylation and molecular weight have a significant effect on the adhesion of chitosan. Although the protonation of chitosan amino groups is dissolved at acidic pH, it is insoluble in most organic solvents. For the reasons mentioned above, chitosan may be an alternative green adhesive [59]. The chemical structure of chitosan is demonstrated in Figure 10.1.

Biomasses such as wood, grass, and grains contain hemicellulose, cellulose, and lignin. Types vary according to chemical composition and

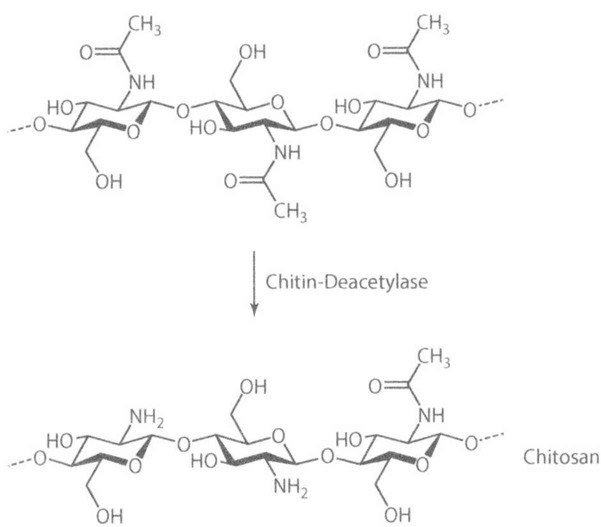

Figure 10.1 Chemical structure of chitosan [60].

structure [54, 61]. Hemicellulose, cellulose microfibril, and pectin are shown in Figure 10.2.

Among the most common forest products in nature are hemicelluloses and gums. Therefore, forestry products and agricultural wastes are used as a natural source of cellulose. Hemicelluloses can take an alternative green adhesive but have problems with water resistance as in starch. However, it is more advantageous than starch since it does not have any food properties. Carob gum (LBG), one of the agricultural wastes, shows high adhesion resistance, water and heat resistance as wood glue [63]. The chemical and dimensional structures of cellulose are displayed in Figure 10.3 and Figure 10.4, respectively [64, 65].

Lignin, one of the most abundant macromolecules, is found in the hard and complex structures of lignocelluloses [32, 66, 67]. Because of this, lignin has recently been investigated [34, 35, 68, 69]. The chemical structure of lignin [70] is demonstrated in Figure 10.5 and 10.6.

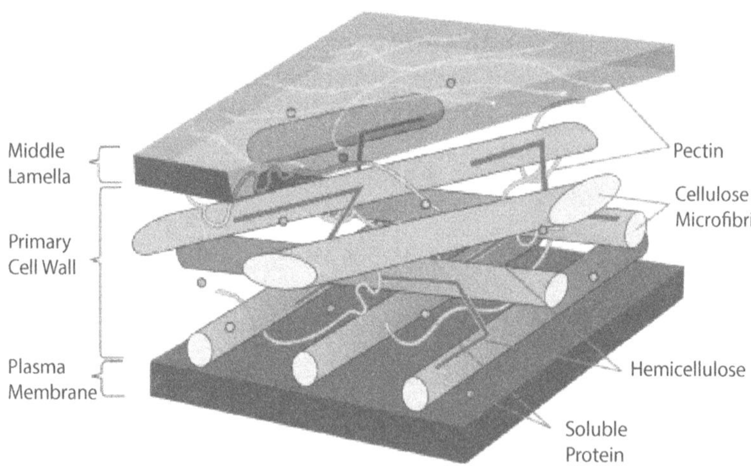

Figure 10.2 Plant cell wall diagram [62].

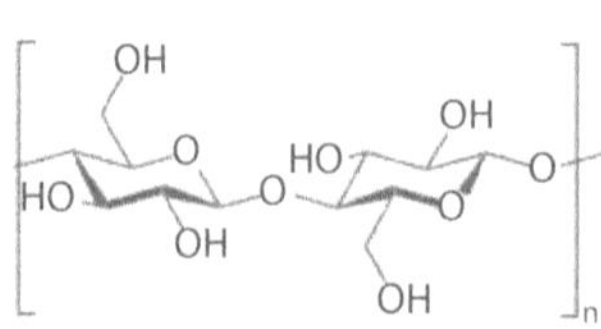

Figure 10.3 Chemical structure of cellulose [64].

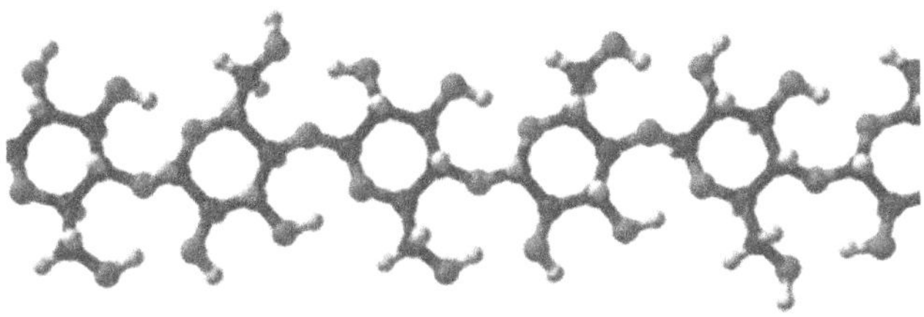

Figure 10.4 Dimensional structure of cellulose [65].

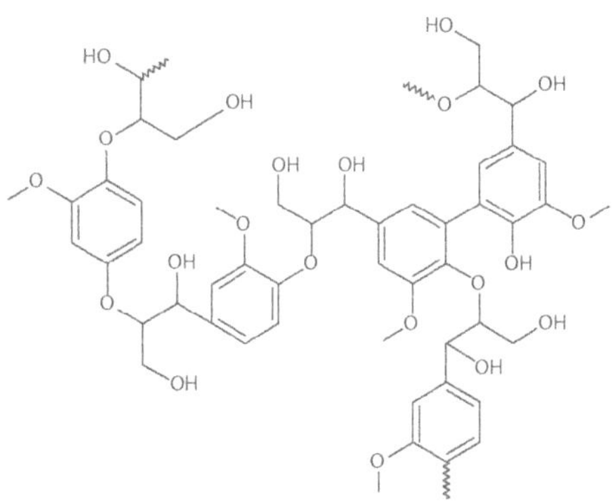

Figure 10.5 Chemical structure of lignin [70].

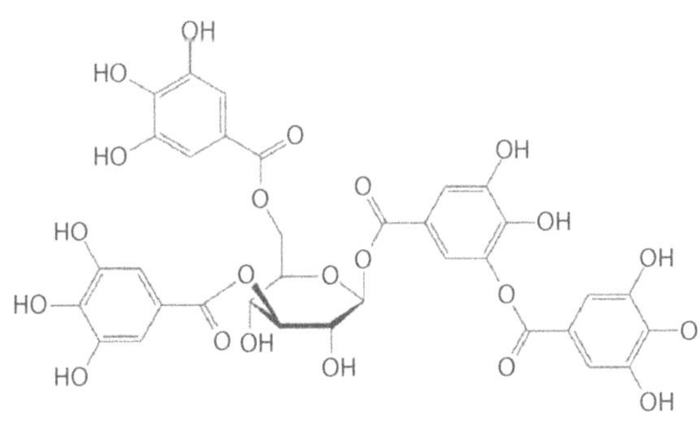

Figure 10.6 Chemical structure of tannin [75].

Since lignin is structurally similar to phenol-formaldehyde, scientists believe that lignin can be used instead of phenol-formaldehyde resin [71].

The polyphenols found naturally in the bark of some trees are called tannins. Examples of these tree species are mimosa and quebracho and pine [72–74].

Similar to lignin [76], because tannins are like water-soluble phenolic compounds. Phenolic resins react with formaldehyde [77].

This book contains critical assessments of green wood flooring adhesives that do not contain formaldehyde and other volatile organic compounds. This section discusses renewable and synthetic adhesives. There are not enough cheap raw materials as an alternative to fossil-based thermosetting wood-flooring adhesives. However, the economic, healthy adhesion strength, and water-resistant green adhesive alternatives required by the wood flooring adhesive industry are evaluated in this section. In addition to the lack of alternative raw materials, toxicological investigations are still not sufficient. Wood, which is an organic material, is known to have formaldehyde emission due to its nature. Serious investigations should be made on the toxicity of adhesives developed as an alternative to formaldehyde resins. Many problems are encountered in the production of green wood flooring adhesive. However, intensive researches in recent years have led to some progress. The most recent developments in this subject will be included in this book chapter. However, due to copyright and other known reasons, formulations of green wood flooring adhesives will not be included in this section.

10.2 Wood Flooring

The wooden floor is any product made of wood that is designed to be used as a floor, whether aesthetic or structural. Wood can come in various styles, cuts, colors, and species because of these reasons wood is a common material for wood flooring. Wooden flooring has aesthetic, hygienic, technological, and economic advantages over alternative materials. There is no other natural upholstery material that can offer aesthetically beautiful possibilities such as wood. Wood flooring makes you feel the closeness and warmth of a raw material that grows naturally aesthetically and gives a relaxation effect to people. Wood flooring can be classified as follows.

10.2.1 Softwood Flooring

Pine, spruce, cedar, fir, juniper, redwood, these tree timbers are used for softwood flooring. An example of pinewood softwood flooring is shown in Figure 10.7.

Figure 10.7 Pinewood flooring.

10.2.2 Hardwood Flooring

As a result of the increasing use of concrete in the construction sector in the world, the wooden floorings used in the interior have gained importance. Oak, walnut (Figure 10.8), beech, mahogany, maple, hickory, balsa, and teak are among the most popular hardwoods used for hardwood flooring. However, hardwood floors are still common and popular. Hardwood floors have a thicker wear surface and can be sanded more times than an engineered wood floor because of its natural structure.

10.2.3 Engineered Wood Flooring

Engineering wood floors consist of two or more veneers of wood bonded together to form aboard to cover indoor surfaces. Engineered wood flooring types are available with different production methods. One of them is wood flooring material which is obtained by adhering lamellas having good surface properties to the upper layer. There are also engineered wood flooring types with different densities. Different compounds are used to increase the fire resistance of wood flooring materials where fibers are bonded with thermosetting adhesives. High fire resistance and non-toxic wood flooring materials are preferred.

10.2.4 Laminate Flooring

The laminate floor is a merged multilayer synthetic floor product together with a lamination process. Laminate floors simulate wood with a layer of photographic sconce under a transparent protective layer.

Figure 10.8 Walnut wood flooring.

10.2.5 Vinyl Flooring

Vinyl floors are widely used today. The development of digital printing technology is one of the most important reasons for this. Thanks to digital technologies on vinyl, the appearance of wood and other natural materials can be given. In vinyl materials, they are fixed to the floor by means of an adhesive. These adhesives may release VOC and formaldehyde emission.

10.2.6 Agricultural Residue Wood Flooring Panels

Eco-design and sustainability principles in the construction sector required the recycling of the materials. In addition, wood flooring factories need raw materials. As a result of the researches conducted to meet the needs of the sector, agricultural wastes have been widely used in plate production in recent years. Barriers to incineration of agricultural wastes are among the most important reasons for the utilization of these wastes. Agricultural waste fibers are compressed using fossil-based or green adhesives and particleboard, medium density fiberboard, or oriented strand boards are produced Agricultural waste boards are made from residues such as poppy husk, corn, and sunflower [1, 78]. The panels produced from agricultural wastes have superior aesthetic properties. The use of agricultural waste in wood flooring can help to preserve the forest presence and our future. In order to produce sustainable structures and adhere to the principles of eco-design, agricultural wastes must be bond with green wood flooring adhesives [78].

The adhesives used in the production of composite panels should not harm human health as well as provide the physical and mechanical requirements of the material. In this regard, the principles of green building and eco-design have been adopted in recent years and regulations have been taken into consideration. The manufacturing stages of poppy husk composite panels are shown in Figure 10.9. Wastes brought to appropriate fiber length and moisture ratio are put into the cold press after being glued. Thin fibers are used in the outer layers to make the surfaces of the composite plate smooth. However, thin fibers may cause a decrease in the mechanical values of the panels. The panel blank is then pressed at a temperature of 160°C–180°C. Produced panels are sized and used as wood flooring by covering or without covering with vinyl or other materials.

10.3 Recent Advances About Green Wood-Flooring Adhesives

The protection of human health and nature is becoming increasingly important. Because of this circumstance, the non-toxic adhesives used for wood flooring became more important. Scientists have made great efforts recently to meet manufacturers' demand for healthy adhesives. The most recent scientific studies for the production of green adhesive are mentioned in this section.

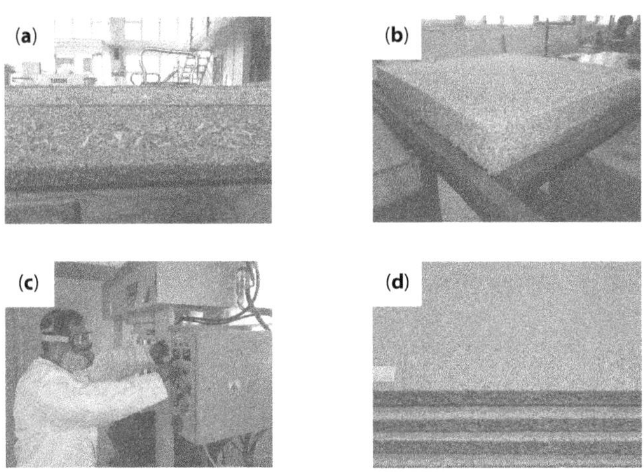

Figure 10.9 Poppy husk based composite panel manufacturing process: (a) cold press, (b) panel blank, (c) hot press, (d) composite panels.

10.3.1 Xylan

Xylan was used in a study to use hemicelluloses as green binders in beech wood adhesives. Commercial hemicellulose was used to minimize natural variations between the research groups. Because such differences can make evaluation difficult by producing ambiguous results. There may be different properties between the purchased beech xylan and the xylan, a byproduct of the pulp. In this study, bonding resistance, heat and water resistance of xylan were investigated. In addition, the dispersion of xylan in water was studied. In this study, trimethylolpropane triacetoacetate, hexa (methoxymethyl) melamine, glycocoll, butanetracarboxylic acid, and citric acid were used as cross-linkers. The comparison of the test results for the dispersions was performed with PVAc, a wood adhesive commonly used in wood flooring. The biggest obstacle to the use of xylan as a wood adhesive is its low waterproofing and low bonding performance. The addition of the above-mentioned dispersing agents or cross-linkers such as hexa (methoxymethyl) melamine to the adhesive yielded good results. Wood flooring coatings bound with xylan adhesive exhibited good water resistance and adhesion strength. Some xylan dispersions exhibited heat and bond strength in accordance with the European Norms EN 14257 and EN 204. The xylan dispersed in the polyvinyl amine adhesive gave results in accordance with the D2 criteria according to EN 204 [9].

10.3.2 Modified Cassava Starch Bioadhesives

In one study, cassava starch was modified with polycarboxylic acid to produce green adhesive. ATR-FTIR studies were conducted to investigate the interaction between green adhesive and substrate. Cassava starch suspensions have been modified with green adhesives. Polycarboxylic acid, citric acid, butane-tetracarboxylic acid, and natural starch were used in this modification. Medium-density fiberboard (MDF) production parameters were determined under laboratory conditions. The developed adhesives showed pseudoplastic behavior. BTCA concentration showed better cross-linking capacity compared to CA. Stress values were 19 and 11 MPA at 80% BTCA and CA concentration, respectively. In this study, it was seen that polycarboxylic acid increased water and adhesion resistance in the panels produced. These results show that green adhesives may replace synthetic adhesives in the near future [79].

10.3.3 High-Efficiency Bioadhesive

Protein-based adhesives do not have sufficient performance due to their poor adhesion and late curing. In this study, it has been tried to develop a green adhesive that can provide energy efficiency by using soy proteins in the laboratory environment. For this purpose, the mineralized adhesive was produced by using soy protein and kenaf fibers. Thermal conductive networks increased the thermal curing rate of SPA-LBN and KF resins. Using this formulation, energy consumption can be reduced by 23%–27% when large scale production is performed. Multifunctional reinforcing additives added to thermoset resin increased the adhesion resistance of adhesives. With this result, a new alternative can be used in the field of wood flooring [80].

10.3.4 Bioadhesive Made From Soy Protein and Polysaccharide

This study was conducted to produce a healthy green adhesive from soy flour. In this study, polysaccharide, which is a water-soluble and cheap material, is combined with the soy protein to produce a new green adhesive. SPI adhesive has good results in terms of thermal and mechanical properties. There is also a decrease in water absorption values. The produced plywood has reached the water-resistance values (0.99 MPa) that can be used indoors. The green adhesive produced in this study can be used in wood flooring. With the current research, a low cost, environmentally friendly adhesive with good water resistance was produced. With the technology used in this study, it is possible to produce green composites by using soybean flour and carbohydrates [38].

10.3.5 Green Cross-Linked Soy Protein Wood Flooring Adhesive

A lot of research is being done on producing adhesive from soy flour. The aim of this study was to produce a two-component soy protein adhesive without adding toxic substances. Polysaccharide used in this study to produce soy protein; obtained from a plant-based, water-soluble soybean. Three kinds of plywood were produced with the adhesive produced. Then, the shear strength of these plywoods was tested. The shear strength results of the produced plywoods were measured as 0.99 MPa. In addition, the water resistance values of the produced samples reached sufficient values for indoor use.

The reason for this development is explained below. Cross-linked soybean soluble polysaccharide and soy protein have formed a network. This web structure is a nested web structure that improves the thermal stability and water resistance of the studied adhesive. The viscosity of the binder was reduced by the addition of CL-SSPS. The adhesive produced thus penetrated the surface of the wood flooring wood thoroughly and provided more adhesion [38].

10.3.6 "Green" Bio-Thermoset Resins Derived From Soy Protein Isolate and Condensed Tannins

Mussel protein has high water resistance. Condensed tannins (CT) have a catechol structure similar to the water-resistance of mussel protein. In this study, CTs were therefore combined with soy protein isolate (SPI) to develop thermoset resin. SPI and CT reactions were catalyzed at different pH values. The chemical structure, thermal decomposition behavior, morphology, and physical properties of the thermosetting resins produced in this study were tested. The resin produced for this purpose was used in the production of plywood. Condensed tannins (CT) have high reactivity and a similar catechol structure. It has been found that the cross-linked structure formed by SPI and CT has improved thermal stability. It was observed that the resin formed a homogeneous fracture surface. Therefore, the water-resistance of the resin increases [81].

10.3.7 Development of Green Adhesives Using Tannins and Lignin for Fiberboard Manufacturing

Today, it is aimed to use natural phenolic tannin and lignin polymers in the production of adhesives instead of petroleum-based adhesives that emit formaldehyde emissions. In this study, tannin extractions and adhesive properties which are an industrial waste of paper mills were evaluated. Liquid extractions in the barks of five different wood species were investigated in the laboratory. Aleppo pine, spruce, fir, sea pine, and eucalyptus are among the tree species used. Aleppo pine was 15% more productive than other tree bark. Sulfide and urea, which are used as liquid additives, were effective in obtaining the tannins in the barks of fir and spruce. Eucalyptus barks were the least tannins obtained. In general, phenol and catechol tannins were found in the bark extracts of Aleppo pine and fir. Adhesion properties of the obtained tannins were also investigated. Liquids from pulp mills are very rich in lignin. Lignin was obtained by precipitating black liquors after acidification. The resulting lignin exhibits interesting adhesion properties after glycosylation. In this study, the adhesion strength of the panels produced with 60% tannin and 40% lignin meets the requirements of European standards [82].

10.3.8 Cottonseed Protein as Wood Adhesives

Recently, soy protein has been commercially used as wood flooring adhesive. In an earlier study, soy protein and cottonseed isolates were used for bonding maple wood coatings. As a result of the research, it was found that cotton seed protein isolate had higher adhesion strength and hot water resistance. This study tested the effects of several acids (amino acids, fatty acids) and several protein modifiers on adhesion, including other molecules with cationic and anionic charges. When acetic acid, butyric acid, glutamic acid, and aspartic acid were added to the cottonseed isolate as particulate acid, it was found to improve adhesion performance. It was found that these acids had no effect on soy protein isolate [83].

10.3.9 Chitosan as an Adhesive

After chitosan alkaline deacetylation, they are obtained from non-plant organisms, crustaceans, and fungi. It is a well-known polysaccharide that has undergone much research in recent years. This is a polysaccharide obtained from unique polycyclic biological sources and soluble in acidic pH. Chitosan is a unique option that can be used in a variety of industrial areas, particularly those requiring biocompatibility. As a result, especially in the medical world, chitosan has become a popular biopolymer. Developments in mechanical properties have also attracted the attention of the industry. This means that chitosan can compete with fossil borne thermosetting adhesives due to its performance as binders. However, studies on the development of biopolymers should be continued [59].

10.3.10 PE-cg-MAH Green Wood Flooring Adhesive

In this study, an adhesive without formaldehyde emission was produced for use in plywood production. This adhesive is produced from *in situ* chlorinating graft copolymerization. The main component of this binder is high-density polyethylene (HDPE) and maleic anhydride (MAH).

MAH is bonded with HDPE (PE-cg-MAH). The best adhesion resistance and hot press temperature of the adhesive in plywood production were investigated. The boiling water test was performed to measure water resistance. Experimental results showed that plywoods applied at 160–165 C and 5 minutes of hot press meet Type I plywood characteristics. The highest shear strength was obtained after the boiling water test when the amount of chlorine was 3% in the glued plates with PE-cg-MAH [84].

References

1. Kubba, S., *Green construction project management and cost oversight*, pp. 71–200, Elsevier Inc., Amsterdam, North Holland, 2010.
2. Brezet, H. and Van Hemel, C., *Ecodesign: A Promising Approach to Sustainable Production and Consumption*, UNEP, Paris, France, 1997.
3. Guidice, F., La Rosa, G., Risitano, A., *Product Design for the Environment: A Life Cycle Approach*, CRC Press e Taylor & Francis Group, Boca Raton, FL, USA, 2006.
4. Cobut, A., Beauregard, R., Blanchet, P., Reducing the environmental footprint of interior wood doors in non-residential buildings e part 2: Eco-design. *J. Cleaner Prod.*, 109, 247–259, 2015.
5. Frihart, C.R., Introduction to special issue, wood adhesives: Past, present, and future. *For. Prod. J.*, 65, 4–8, 2015.
6. Vick, C.B., Adhesive bonding of wood materials, in: *Wood handbook: Wood as an engineering material*, Laboratory FP, (Ed.), United States Department of Agriculture Forest Service, Madison, Wisconsin, USA, 1999.
7. Stoeckel, F., Konnerth, J., Gindl-Altmutter, W., Mechanical properties of adhesives for bonding wood—A review. *Int. J. Adhes. Adhes.*, 45, 32–41, 2013.
8. Dunky, M. and Niemz, P., *Holzwerkstoffe und Leime: Technologie und Einflussfaktoren*, Springer Berlin Heidelberg, Berlin, Germany, 2002.
9. Norström, E., Fogelström, L., Nordqvist, P., Khabbaz, F., Malmström, E., Xylan – A green binder for wood adhesives. *Eur. Polym. J.*, 67, 483–493, 2015, http://dx.doi.org/10.1016/j.eurpolymj.2015.02.021.
10. Pizzi, A., Synthetic adhesives for wood panels: Chemistry and technology—A critical review. *Rev. Adhes. Adhes.*, 2, 85–126, 2014.
11. Zeppenfeld, G. and Grunwald, D., *Klebstoffe in der Holz- und Mobelindustrie*, DRWVerlag, Leinfelden-Echterdingen, Germany, 2005.
12. Solt, P., Konnerth, J., Gindl-Altmutter, W., Kantner, W., Moser, J., Mitter, R., Herwijnen, H.W.G., Technological performance of formaldehyde-free adhesive alternatives for particleboard industry. *Int. J. Adhes. Adhes.*, 94, 99–131, 2019, https://doi.org/10.1016/j.ijadhadh.2019.04.007.

13. Whitfield, R.M., Brown, F.C., Low, R., *Socio-economic benefits of formaldehyde to the European Union (EU 25) and Norway*, Global Insight, Lexington, M.A, 2007.
14. Roffael, E., Formaldehyde release from wood-based panels—A review. *Holz Roh Werkst.*, 47, 41–5, 1989.
15. Roffael, E., Kraft, R., Behen, C., Formaldehydabgabe von Holzwerkstoffen von gestern nach ubermorgen, Teil 1. *Holztechnologie*, 49, 48–52, 2008.
16. Salthammer, T., Mentese, S., Marutzky, R., Formaldehyde in the indoor environment. *Chem. Rev.*, 110, 2536–7, 2010.
17. Wolcott, J.J., Motter, W.K., Daisy, N.K., Tenhaeff, S.C., Detlefsen, W.D., Investigation of variables affecting hot-press formaldehyde and methanol emissions during laboratory production of urea-formaldehyde-bonded particleboard. *For. Prod. J.*, 46, 62–8, 1996.
18. Gann, M., Emissionsarme Harze fur technische Anwendung, Teil 1: Aktuelle Entwicklung bei Formaldehydemssionstandards. *IBO Mag.*, 3, 26–9, 2009.
19. Schafer, M. and Roffael, E., On the formaldehyde release of wood. *Holz Roh Werkst.*, 58, 259–64, 2000.
20. Salem, M.Z.M. and Bohm, M., Understanding of formaldehyde emissions from solid wood: An overview. *Bioresources*, 8, 4775–90, 2013.
21. Risholm-Sundman, M., Larsen, A., Vestin, E., Weibull, A., Formaldehyde emission-comparison of different standard methods. *Atmos. Envron.*, 41, 3193–202, 2007.
22. Hauptmann, M., Lubin, J.H., Stewart, P.A., Hayes, R.B., Blair, A., Mortality from solid cancers among workers in formaldehyde industries. *Am. J. Epidemiol.*, 159, 1117–1130, 2004.
23. Boran, S., Usta, M., Gumuskaya, E., Decreasing formaldehyde emission from medium density fiberboard panels produced by adding different amine com pounds to urea formaldehyde resin. *Int. J. Adhes. Adhes.*, 31, 674–8, 2011.
24. Costa, N.A., Pereira, J., Ferra., J., Cruz, P., Martins, J., Magalhes, F.D. *et al.*, Scavengers for achieving zero formaldehyde emission of wood-based panels. *Wood Sci. Technol.*, 47, 1261–72, 2013.
25. Kim, S., Kim, H.J., Kim, H.S., Lee, H.H., Effect of bio-scavengers on the curing behavior and bonding properties of melamine-formaldehyde resins. *Macromol. Mater. Eng.*, 291, 1027–34, 2006.
26. Fliedner, E., Heep, W., van Herwijnen, H.W.G., Verwendung nachwachsender Rohstoffe in Bindemitteln fur Holzwerkstoffe. *Chem. Ing. Tech.*, 82, 1161–8, 2010.
27. European Patent Office Granted Patent number: EP1318000B1, Krug, D., Sirch, H.J., Lang, J., Heep, W., Hopcke, R., Binder for manufacturing and binding of wood and timber products, 2006.
28. Granted Patent number: WO2009077571, Van Herwijnen, H.W.G., Heep, W., Krug, D., Weber, A., Hohling, A., Schultze, M., Binder for materials based on wood chips and/or wood fibers, method for the production of said binder, and molded article, 2009.

29. Myers, G.E., *Mechanisms of formaldehyde release from bonded wood products, Formaldehyde Release from Wood Products*, vol. 87-106, American Chemical Society, Washington, USA, 1986.
30. Norström, E., Demircan, D., Fogelström, L., Khabbaz, F., Malmström, E., Green Binders for Wood Adhesives, London, UK, pp. 49–71, Intechopen, 2018, http://dx.doi.org/10.5772/intechopen.72072.
31. Gomez-Bueso, J. and Haupt, R., Wood composite adhesives, in: *Phenolic resins: A century of progress*, L. Pilato (Ed.), Springer Verlag, Heidelberg, 2010.
32. Frihart, C.R., Wood adhesion and adhesives, in: *Handbook of Wood Chemistry and Wood Composites*, R.M. Rowell (Ed.), pp. 215–278, CRS Press, Florida, 2005.
33. Vnucec, D., Kutnar, A., Gorsek, A., Soy-based adhesives for wood-bonding—A review. *J. Adhes. Sci. Technol.*, 31, 8, 910–931, 2017.
34. Hemmilä, V., Adamopoulos, S., Karlsson, O., Kumar, A., Development of sustainable bioadhesives for engineered wood panels—A review. *RSC Adv.*, 7, 61, 38604–38630, 2017.
35. Ferdosian, F., Pan, Z., Gao, G., Zhao, B., Bio-based adhesives and evaluation for wood composites application. *Polymers (Basel, Switz)*, 7, 2, 70, 2019.
36. Uses of Soybean, http://ncsoy.org/media-resources/uses-of-soybeans/NC Soybean Producer Association, 2019.
37. Zeng, N., Xie, J., Ding, C., Properties of the soy protein isolate/PVAc latex blend adhesives, in: *Advanced Materials Research*, vol. 550–553, pp. 1103–7, (Pt. 2, Advances in Chemical Engineering II), Durnten, Zurich, Switzerland, 2012.
38. Yuan, C., Chen, M., Luo, J., Li, X., Gao, Q., Li, J., A novel water-based process produces ecofriendly bio-adhesive made from green cross-linked soybean soluble polysaccharide and soy protein. *Carbohydr. Polym.*, 169, 417–425, 2017.
39. Luo, J., Luo, J., Yuan, C., Zhang, W., Li, J., Gao, Q. *et al.*, An eco-friendly wood adhesive from soy protein and lignin: Performance properties. *RSC Adv.*, 5, 122, 100849–100855, 2015.
40. Pradyawong, S., Qi, G., Li, N., Sun, X.S., Wang, D., Adhesion properties of soy protein adhesives enhanced by biomass lignin. *Int. J. Adhes. Adhes.*, 75, 66–73, 2017.
41. Xin, J., Zhang, P., Wolcott, M.P., Zhang, J., Hiscox, W.C., Zhang, X., A novel and formaldehydefree preparation method for lignin amine and its enhancement for soy protein adhesive. *J. Polym. Environ.*, 25, 3, 599–605, 2017.
42. Cheng, H.N., Ford, C., Dowd, M.K., He, Z., Soy and cottonseed protein blends as wood adhesives. *Ind. Crops Prod.*, 85, 324–330, 2016.
43. D'Amico, S., Mueller, U., Berghofer, E., Effect of hydrolysis and denaturation of wheat gluten on adhesive bond strength of wood joints. *J. Appl. Polym. Sci.*, 129, 2429–2434, 2013.

44. El-Wakil, N.A., Abou-Zeid, R.E., Fahmy, Y., Mohamed, A.Y., Modified wheat gluten as a binder in particleboard made from reed. *J. Appl. Polym. Sci.*, 106, 3592–3599, 2007.
45. Khosravi, S., Nordqvist, P., Khabbaz, F., Johansson, M., Protein-based adhesives for particleboards-effect of application process. *Ind. Crops Prod.*, 34, 1509–1515, 2011.
46. Lei, H., Pizzi, A., Navarrete, P., Rigolet, S., Redl, A., Wagner, A., Gluten protein adhesives for wood panels. *J. Adhes. Sci. Technol.*, 24, 583–1596, 2010.
47. Nordqvist, P., Khabbaz, F., Malmström, E., Comparing bond strength and water resistance of alkali-modified soy protein isolate and wheat gluten adhesives. *Int. J. Adhes. Adhes.*, 30, 72–79, 2010.
48. Nordqvist, P., Lawther, M., Malmström, E., Khabbaz, F., Adhesive properties of wheat gluten after enzymatic hydrolysis or heat treatment—A comparative study. *Ind. Crops Prod.*, 38, 139–145, 2012.
49. Nordqvist, P., Nordgren, N., Khabbaz, F., Malmström, E., Plant proteins as wood adhesives: Bonding performance at the macro- and nanoscale. *Ind. Crops Prod.*, 44, 246–252, 2013.
50. Nordqvist, P., Thedjil, D., Khosravi, S., Lawther, M., Malmström, E., Khabbaz, F., Wheat gluten fractions as wood adhesives-glutenins versus gliadins. *J. Appl. Polym. Sci.*, 123, 1530–1538, 2012.
51. Khosravi, S., Nordqvist, P., Khabbaz, F., Oehman, C., Bjurhager, I., Johansson, M., Wetting and film formation of wheat gluten dispersions applied to wood substrates as particle board adhesives. *Eur. Polym. J.*, 67, 476–482, 2015.
52. Khosravi, S., Khabbaz, F., Nordqvist, P., Johansson, M., Wheat-gluten-based adhesives for particle boards: Effect of crosslinking agents. *Macromol. Mater. Eng.*, 299, 116 124, 2014.
53. Khosravi, S., Khabbaz, F., Nordqvist, P., Johansson, M., Protein-based adhesives for particleboards. *Ind. Crops Prod.*, 32, 275–283, 2010.
54. Patel, A.K., Mathias, J.D., Michaud, P., Polysaccharides as adhesives: A critical review. *Rev. Adhes. Adhes.*, 1, 3, 312–345, 2013.
55. Gadhave, R.V., Mahanwar, P.A., Gadekar, P.T., Starch-based adhesives for wood/wood composite bonding: Review. *Open J. Polym. Chem.*, 7, 2, 19–32, 2017.
56. Imam, S.H., Gordon, S.H., Mao, L., Chen, L., Environmentally friendly wood adhesive from a renewable plant polymer: Characteristics and optimization. *Polym. Degrad. Stab.*, 73, 529–533, 2001.
57. Wang, Z., Li, Z., Gu, Z., Hong, Y., Cheng, L., Preparation, characterization and properties of starch-based wood adhesive. *Carbohydr. Polym.*, 88, 699–706, 2012.
58. Zargar, V., Asghari, M., Dashti, A., A review on chitin and chitosan polymers: Structure, chemistry, solubility, derivatives, and applications. *ChemBioEng Rev.*, 2, 3, 204–226, 2015.
59. Mati-Baouche, N., Elchinger, P.H., de Baynast, H., Pierre, G., Delattre, C., Michaud, P., Chitosan as an adhesive. *Eur. Polym. J.*, 60, 198–212, 2014.

60. Chemical structure of chitosan, https://en.wikipedia.org/wiki/Chitosan, Wikipedia, 2019.
61. Salam, A., Venditti, R.A., Pawlak, J.J., El-Tahlawy, K., Crosslinked hemicellulose citrate-chitosan aerogel foams. *Carbohydr. Polym.*, 84, 1221–1229, 2011.
62. Plant cell wall diagram, https://en.wikipedia.org/wiki/Cell_wall, Wikipedia, 2019.
63. Norström, E., Fogelström, L., Nordqvist, P., Khabbaz, F., Malmström, E., Gum dispersions as environmentally friendly wood adhesives. *Ind. Crops Prod.*, 52, 736–744, 2014.
64. Chemical structure of cellulose, https://en.wikipedia.org/wiki/Cellulose, Wikipedia, 2019.
65. Dimensional structure of cellulose, https://en.wikipedia.org/wiki/Cellulose, Wikipedia, 2019.
66. Younesi-Kordkheili, H., Improving physical and mechanical properties of new lignin-urea-glyoxal resin by nanoclay. *Eur. J. Wood Wood Prod.*, 6, 1–7, 2017.
67. Fu, D., Mazza, G., Tamaki, Y., Lignin extraction from straw by ionic liquids and enzymatic hydrolysis of the cellulosic residues. *J. Agric. Food Chem.*, 58, 5, 2915–2922, 2010.
68. Ghaffar, S.H. and Fan, M., Lignin in straw and its applications as an adhesive. *Int. J. Adhes. Adhes.*, 48, 92–101, 2014.
69. Pizzi, A., Recent developments in eco-efficient bio-based adhesives for wood bonding: Opportunities and issues. *J. Adhes. Sci. Technol.*, 20, 829–846, 2006.
70. Chemical structure of lignin, https://en.wikipedia.org/wiki/Lignin, Wikipedia, 2019.
71. Ibrahim, M.N.M., Zakaria, N., Sipaut, C.S., Sulaiman, O., Hashim, R., Chemical and thermal properties of lignins from oil palm biomass as a substitute for phenol in a phenol formaldehyde resin production. *Carbohydr. Polym.*, 86, 1, 112–119, 2011.
72. Kim, S., Environment-friendly adhesives for surface bonding of wood-based flooring using natural tannin to reduce formaldehyde and TVOC emission. *Bioresour. Technol.*, 100, 2, 744–748, 2009.
73. Pinto, P.C., Sousa, G., Crispim, F., Silvestre, A.J., Neto, C.P., Eucalyptus globulus bark as source of tannin extracts for application in leather industry. *ACS Sustainable Chem. Eng.*, 1, 8, 950–955, 2013.
74. Fechtal, M. and Riedl, B., Use of eucalyptus and Acacia mollissima bark extract-formaldehyde adhesives in particleboard manufacture. *Holzforschung*, 47, 4, 349–357, 1993.
75. Chemical Structure of Tannin, https://en.wikipedia.org/wiki/Tannin, Wikipedia, 2019.
76. Faris, A.H., Rahim, A.A., Ibrahim, M.N.M., Hussin, M.H., Alkurdi, A.M., Salehabadi, A., Investigation of oil palm based Kraft and auto-catalyzed

organosolv lignin susceptibility as a green wood adhesives. *Int. J. Adhes. Adhes.*, 74, 115–122, 2017.
77. Faris, A.H., Ibrahim, M.N.M., Rahim, A.A., Preparation and characterization of green adhesives using modified tannin and hyperbranched poly(amine-ester). *Int. J. Adhes. Adhes.*, 71, 39–47, 2016.
78. Keskin, H., Kucuktuvek, M., Guru, M., The potential of poppy (Papaver somniferum Linnaeus) husk for manufacturing wood-based particleboards. *Constr. Build. Mater.*, 95, 224–231, 2015.
79. Monroy, Y., Sandra Rivero, S., Garcia, M.A., Sustainable panels design based on modified cassava starch bioadhesives and wood processing byproducts. *Ind. Crops Prod.*, 137, 171–179, 2019, https://doi.org/10.1016/j.indcrop.2019.04.062.
80. Wang, Z., Zhao, S., Kang, H., Zhang, W., Li, J., Zhang, S., Huang, A., Reduction of energy consumption of green plywood production by implementing high-efficiency thermal conductive bio-adhesive: Assessment from pilot-scaled application. *J. Cleaner Prod.*, 210, 1366–1375, 2019, https://doi.org/10.1016/j.jclepro.2018.10.326.
81. Liu, C., Zhang, Y., Li, X., Luo, J., Gao, Q., Li, J., "Green" bio-thermoset resins derived from soy protein isolate and condensed tannins. *Ind. Crops Prod.*, 108, 363–70, 2017, http://dx.doi.org/10.1016/j.indcrop.2017.06.057.
82. Bertaud, F., Tapin-Lingua, S., Pizzi, A., Navarrete, P., Petit-Conil, M., Development of green adhesives for fibreboard manufacturing, using tannins and lignin from pulp mill residues. *Cell. Chem. Technol.*, 46, 7–8, 449–455, 2012.
83. Cheng, H.N., Ford, C., Dowd, M.K., He, Z., Use of additives to enhance the properties of cottonseed protein as wood adhesives. *Int. J. Adhes. Adhes.*, 68, 156–160, 2016, http://dx.doi.org/10.1016/j.ijadhadh.2016.02.012.
84. Tang, L., Zhang, Z.G., Qi, J., Zhao, J.R., Feng, Y., The preparation and application of a new formaldehyde-free adhesive for plywood. *Int. J. Adhes. Adhes.*, 31, 6, 507–512, 2011, https://www.sciencedirect.com/science/article/pii/S0143749611000546

11

Synthetic Binders for Polymer Division

Sathish Kumar Palaniappan[1]*, Moganapriya Chinnasamy[2], Rajasekar Rathanasamy[2] and Samir Kumar Pal[1]

[1]Department of Mining Engineering, Indian Institute of Technology Kharagpur, West Bengal, India
[2]Department of Mechanical Engineering, Kongu Engineering College, Tamil Nadu, India

Abstract

Synthetic binders are special adhering materials manufactured from pure polymers, polymeric blends, resins, and oils. The chapter focuses on the basic characteristics, classification, production, benefits, properties, application of synthetic adhesives, and resins for bonding glass, metals, ceramics, wood, rubber, foil, paper, leather, cloth, plastic films, and some structural materials. Most synthetic polymeric binders such as thermosets, thermoplastics and elastomers or combination of these types are dissolvable-based scatterings or unadulterated solids to have a decent blend of properties and are commonly more grounded over more extensive temperature ranges compared to natural binders. The survey has been riveted on the synthetic binders such as polyesters, epoxy, acrylic, phenol formaldehyde, melamine, polyimides, phenolic neoprene, silicones, furan, polyurethane, polyolefins, etc. Attention has been provided on the technical aspects such as chemical properties, thermal properties, and mechanical properties like impact strength, creep strength, shear strength, toughness, elongation, peel strength, etc., that make the binders deserved for bonding applications. Appropriate assortment of different fillers, hardeners, modifiers, and resins are the primary factors that lead to desired properties of the synthetic binders for a suitable application. It is observed that synthetic binders have steady change in their structure, binding characteristics, mechanical strength, chemical resistance, and thermal properties compared to nature born binders. These few advantages of synthetic polymers with desirable properties find applications superior to natural binders.

**Corresponding author*: sathishiitkgp@gmail.com

Inamuddin, Rajender Boddula, Mohd Imran Ahamed and Abdullah M. Asiri (eds.) Green Adhesives: Preparations, Properties and Applications, (227–272) © 2020 Scrivener Publishing LLC

Keywords: Synthetic binders, mechanical properties, elastomeric adhesives, bonding, epoxy resins, polyurethanes, phenol formaldehyde resins, melamine resins

List of Abbreviations

UV	ultraviolet
PI	polyimides
PBI	polybenzimidazoles
wt.%	weight percentage
°C	degree celsius
°F	degree Fahrenheit
ASTM	American Society for Testing and Materials
HCl	hydrochloric
CO	carbon monoxide
PPO	polyphenylene oxide
KOH/g	potassium hydroxide per gram
mg	milligram
g/cc	gram per cubic centimeter
m^2/g	meter square per gram
mm	millimeter
cm^3/g	cubic centimeter per gram
PET	polyethylene terephthalate
TEM	transmission electron microscopy
HBP	hyper spread polymers
DGEBA	diglycidyl ether of bisphenol A
TGDDM	tetraglycidyl-4,4′-diaminodiphenylmethane
LC	liquid crystals
CTBN	carboxy-ended butadiene/acrylonitrile
NBR	nitrile rubber
PMMA	poly (methyl methacrylate)
F-PEK	fluorinated poly (aryl ether ketone)
PEG	polyethylene glycol
TEOS	tetraethoxysilane
PEEK	poly (ether ether ketone)
DMA	dynamic mechanical analysis
FTIR	Fourier transform infra-red spectroscopy
SEM	scanning electron microscopy
HON	high ortho novolak resins
GPN	general-purpose novolak resins

HPN	high para novolak resins
CHP	cumenehydroperoxide
MPa	megapascal
g/ml	gram per milliliter
g/mol	gram per mol
nm	nanometer
PF	phenol based formaldehyde
VOCs	volatile organic compounds
CIBA	Chemische Industrie Basel
FA	furfuryl alcohol
HMF	hydroxymethylfurfural
DA	Diels-Alder response
THF	tetrahydrofuran
C-NMR	carbon-13 nuclear magnetic resonance
HEC	2-hydroxy-ethyl-cellulose
MA	methacrylamide
ST	styrene
DVB	divinylbenzene
TBA	tetrabutylammonium perchlorate
Hep	heptane
Tol	toluene
MLAU	methacrylated lauric acid
MOA	oleic acid glycidyl methacrylate

11.1 Introduction

The early development of binders was more empirical than technical, cause effect analysis, and understanding. An increasing cause of spread and combination of glues, gums, resins, high polymers, and many special additives are being used as adhesives with emerging emphasis in their efficient manufacture and effective formulations. Adhesion science aims in understanding the materials and properties related with formation of the interfaces, variations in the interfaces with time and events allied with failure of the interfaces. Two similar or dissimilar surfaces, which are placed together through surface forces of attraction and mechanical interlocking is generally described as adhesion. It comprises an adherent, organic resin or polymer adhesive, and intervening interface. Polymer adhesive joints may include one or more interfaces. Polymers mainly called as macromolecules joined by a chain of monomers and the process are called as polymerization. The internal structure of molecular chains defines the specific properties of polymers. Polymer adhesives are mainly available in liquid state, semi-liquid state, and/or viscoelastic phase, and the three techniques to harden the adhesives from liquid to solid state are:

- Dissolved in water or solvents and allow it to dry, i.e., drying adhesives
- Melted and allow it to solidify upon cooling, i.e., hot melts
- Use of cure precursors for phase transformation

The common curing processes of polymer adhesive are (i) mixing, (ii) heating, (iii) adhesives for curing, (iv) moisture detection, and (v) kept in absence of oxygen. All the above mentioned process can also be combined depending on their application [1]. Low viscous adhesives got evaporated during the process or at the initial stage, which is a best suited solvent for thermosetting materials. The various optical and ultraviolet (UV) properties of synthetic polymer adhesives were extensively studied in detail [2–5]. They concluded that curing process initiated through UV light and continues even after removing the source. The pressure sensitive adhesives were also analyzed for optical applications. High viscous polymers generally deform and flow slowly to provide low bonding strength finds its application in pressure sensitive adhesives. In industries, thermal curing, UV curing, solvents drying through thermal or UV source, and two-component curing are the majorly used hardening methods for polymeric adhesives [6].

11.2 Classification of Adhesives Based on Its Chemical Properties

Thermoset, thermoplastic, and elastomeric are the three major classifications of synthetic adhesives.

11.2.1 Thermoset Adhesives

The thermoset adhesives can't be warmed and softened after the underlying fix. Relieving happens by synthetic responses at elevated temperature based upon the kind of glues. Some adhesives need significant weight, whereas others involve just pressure. Solvents are in some cases supplementary to encourage its usage. These adhesives are generally accessible as dissolvable free fluids, glues, and solids. Thermoset cements are distinguished as one and two section frameworks. The one-section frameworks generally require curing at higher temperature and have a restricted timeframe of realistic usability. The two-section frameworks have longer timeframes of realistic usability and can be relieved gradually at room temperature or to some degree quicker at decently elevated temperatures. An impediment is that the requirement for cautious metering and blending to ensure that the endorsed extents is mixed and the resultant blend is homogeneous. When the adhesives are blended, the service life is restricted.

Thermoset glues, when relieved are thickly cross-connected, their protection from warmth and solvents is great and they show minimal adaptable disfigurement under raised temperatures. Generally, temperatures of 93°C–260°C is the withstanding capacity for cross-linked bonds which possess fair peel strength, which finds its application in stressed joints at elevated temperatures. Some of the thermoset adhesives bonded materials have its own structural applications because of usage in elevated load assemblies and harsh service circumstances. Thermoset adhesives set an outcome for developing atomic chains to deliver an inflexible cross-linked structure. Additionally, more epoxy gums are included, which are probably the most broadly utilized adhesives. Polyesters, polyimides, epoxy-based resins, polysulfide, furan, polysulphides, acrylics, redux adhesive, fluoropolymer, and silicones are the commonly used few diverse thermoset adhesives.

11.2.2 Thermoplastic Adhesives

Thermoplastic materials generally avoid cross-linking at the time of curing process and they can be dissolved without any change in their properties. It is a solitary segment framework that solidifies after cooling from a liquefy state or by dissipation of a dissolvable. Wood pastes are usually utilized thermoplastic emulsions in household applications, which solidifies through emulsion process, i.e., dissipation of water. These are not normally prescribed for use at a temperature over 66°C, in spite of the fact that they can be utilized around 90°C in certain prescribed applications. These resources have reduced jerk opposition and reasonable strip quality. It finds own application in stressed joints, overlaps, stiffeners, and in cap design. The most common non-metallic materials bonded using thermoplastic adhesives are paper, wood, cowhide, and plastics. Inspite of a few hot-liquefy adhesives, these are not commonly utilized for auxiliary purposes.

11.2.3 Adhesive Blends

These adhesives were formulated by combination (blends) of two distinctive compounds, i.e., thermoset, thermoplastic, or else elastomeric gathering. The thermoset adhesives were picked because of its huge strength which makes the mix harder, progressively impervious to effect, and increasingly adaptable. The blends exploit the most significant properties of every component. It is usually accessible as dissolvable-based arrangements and as upheld/unsupported films. Aside from a few epoxy blends, thermal heat and stress are normally essential for curing. Nearly all blend adhesives are dissolvable related scatterings or normally 100% solids. It has decent mix characteristics and is commonly more grounded over more extensive temperature series compared to many different adhesives. These are likewise called basic adhesives utilized for the most military applications. The holding materials incorporate metals, earthenware production, and glass which discovers essentially for elevated power and temperatures.

11.3 Adhesives Characteristics

The general classifications of adhesives are mixes, sweet-smelling polymer, deferred tack, conductive, rubber-based, and hot-soften adhesives, tape adhesives, film and inorganic pastes, adhesives based on water and microencapsulated materials, dissolvable frameworks, thermoplastic, and thermosetting resin adhesives. For all intents, generally, these are produced from plant or creature material proceeding from the decade. The fundamental modules incorporate paste through creature bones, paste obtained from fish, and adhesives developed by vegetables. The advancement within natural science and expansion sought prompted improvements in engineered mixes starting with phenol formaldehyde. The progress in polymerization and glue science continued in a close parallel manner.

Numerous quantities of acrylic resins are utilized for holding calfskin, material, plastics, metal foils, now and again. The acrylic monomers for the most part utilized in adhesives are acrylamide, methyl acrylate, methacrylic corrosive, acrylonitrile, acrylic corrosive, and ethyl acrylate. Mainly, these are dissolvable inside the natural solutions which provided in a similar way as other dissolvable based frameworks. Polymers generally dissolve in monomers. Exactly at the time of impetus addition, monomers that tend to polymerize gives great clinging to plastic and to glass surfaces of comparative structure. An assortment of acrylic copolymers is set up by emulsion polymerization [7]. Various acrylic adhesives called "modified acrylics," "receptive adhesives," "responsive fluid adhesives", and "second-age acrylics" are accessible throughout the years. Basically, polymerization occurred in the paste line and turns out to be a vital piece of a cement compounds. Major classifications of adhesives are epoxy-polysulfide adhesives, polyurethane adhesives, acrylic-based adhesives, epoxy adhesives, polyolefin adhesives and silicone adhesives, epoxy-phenolic adhesives and cyanoacrylate adhesives. The method of production, desired physical, chemical and mechanical properties, advantages, and their end applications vary based on their material characteristics.

11.4 Adhesives Classification Based on Its Function

11.4.1 Permanent Adhesives

Generally, these adhesives are ordinarily determined for general purpose applications, which normally employs improved holding qualities and withstand most natural circumstances. The execution mainly relies upon the connected surface. Some of the permanent adhesives based on their characteristics are as follows:

- *Permanent, Acrylic:* It has excellent bonding to painted metal, paper and high surface vitality plastics, glass, not for wax containers, permeable surfaces, and finished layers.
- *Permanent, Aggressive:* Elastic based lasting or altered acrylic cement is forceful and is hard to expel. They execute fit on folded, painted metal, paper, glass, and additionally troublesome substrates or testing conditions.
- *Permanent, Block Out:* High tack perpetual glue with an uncommon misty obstruction which hindrance coated layers.
- *Permanent, All Temperature:* Uncommon glue applied to surfaces at temperatures of −20°F, which renders great starting tack and elevated extreme bonding. It acts healthy on minimum surface vitality plastics which generally have great permanency with temperature ranges from −65°F to 160°F.
- *Permanent, Oblique Food Contact:* Cement intended towards following new products of the soil with palatable skins, which provides better tack and grip.
- *Permanent, High Heat Resistance:* Superior and acrylic-based glue exhibit high extreme strip strengths with 300°F–350°F

as service temperature. The mixes by means of polyimide along with polyvinyl fluoride film utilized within the developing printed circuit boards.

- *Permanent, Repulpable:* Dispersible resin provides superb attach and grip to folded and paper substrates, whereas non-polar substrates have moderate grip.
- *Permanent, Rubber Based:* It has high introductory tack and high extreme grip with +25°F as application temperature and −65°F to +220°F as operating ranges. It performs well on bended surfaces, paper, painted metal, glass, and bundling film.
- *Permanent, Textile:* Developed explicitly meant for sock band in addition to other textile (excluding silk fabrics) label applications.
- *Permanent, Tire:* Perpetual glue structured explicitly for name applications with high beginning tack, shear in addition to extreme bonding to provide incredible execution on hard to label substrates like wood as well as floor cover.
- *Permanent, Water-Dissolvable:* It is permanent glue, removable in around 100°F of water which has great starting tack, extreme bonding which enables customers to expel the label utilizing warm water. It is additionally powerful, when labels were twisted to compartments.
- *Semi-Permanent/Temporary, Freezer:* Cement with great beginning tack and extreme grip at temperatures as minimum as −20°F applied on the majority at nominal room temperature or more.

11.4.2 Removable Adhesives

Whenever evacuated, the label resolves to disconnect in one piece parting practically no cement buildup on it. Each detachable application involves exhaustive testing that the cement and surface are very good in compatibility. Normally, removable glue is prescribed just a momentary life is connected to a surface.

- *Removable, Frozen Temperature:* It is a general use removable acrylic cement containing great tack and grip execution by means of prevalent removability even in lesser temperatures, which is around +5°F.

- *Removable, Elevated Initial Tack:* It is also removable cement with maximum initial tack, inward potential and steady bonding expels from the layers.
- *Removable, Laser-based:* Extraordinary cements readily comprising great tack and grip properties with unrivaled removability and compatible to laser printer.
- *Removable, Long Term:* Removable cement developed with specific highlights like moderate tack, removability, etc. It is additionally accessible in ultra-removable evaluation.
- *Removable, Special:* Glue with special-less tack, grip, and shear characteristics have −20°F to +250°F as an operating temperature.

11.4.3 Repositionable Adhesives

Like removable adhesives, this enables a mark to be expelled from a surface and connected to the equivalent or diverse layers without trailing its adequacy or parting residue. Generally, this element is attractive in circumstances where labels have been twisted. More adhesion is alluring and it should be evacuated and reapplied. Adhesives in the midst of low or moderate introductory tack fill this function. Presently, there are various removable adhesives which have moderate starting tack and great bond to meet all requirements for this reason. Adhesive testing is extremely suggested.

11.4.4 Blended Adhesives

Mixed adhesives are significant as like structural adhesives particularly during metal attachment. So, they include a thermoset and thermoplastic polymer, i.e., rubbers. Insignificantly, every segment has adhesive properties without any input, as a whole the conjoint framework shapes a more grounded and increasingly adaptable adhesive. Hence, the two-polymer frameworks were especially fruitful as like film in addition to tape-based adhesives.

The characteristics of every segment polymer are altered with adding together, perhaps expanding thermal conflict and lessening of other. So, strength of one might be expanded via sacrificing other's flexibility. In this manner, these are conceivable to detail an assortment of gums by means of a wide scope of attributes by essentially unreliable proportion of polymers. In majorly utilized two-polymer-based adhesives, thermoset segment is a phenolic compound. These are commonly good, in spite of the fact that not effectively miscible with various thermoplastic polymers. Especially great similarity is exhibited between ordinary liquor dissolvable phenolic saps, acetals, and polyvinyl esters. However, epoxies are always significant in two-polymer-based adhesive frameworks. Hence, the mainly significant thermoplastic segments are polyvinyl acetals and manufactured elastic, especially nitrile elastic dissolvable nylons were additionally a significant class. However, five of the mainly significant two-polymer-based adhesives utilized in application of films and tapes incorporate nylon-epoxies, vinyl-phenolics, nitrile-phenolics, elastomer-epoxies, and epoxy-phenolics.

11.4.5 Anaerobic Adhesives

The adhesive utilized is acrylate corrosive diesters (polyester-acrylic) additionally been advanced for use as a sealant. It is basically monomeric meager liquids which polymerize to shape an extreme plastic bond between intently fitting metal joints. Therefore, metal surfaces quicken the polymerization process without air. So, these materials will bond every single normal metal, glass, earthenware production and thermoset plastics to one another. However, polymerization process is basically a radical free expansion process. Thus, the most significant use of anaerobic gums as liquid locks washers mainly for screws and fasteners. In view of solid infiltrating capacity, they are capable to connect previously or after gathering. The overall torque for the most grounded evaluations is ordinarily more prominent compared to locknuts in addition to lock screws. Somehow, speed of the cure is expanded to a great extent reliant on parts joined. So, there are basically three fundamental cure-speed categories like quick, medium, and moderate at normal temperature devoid of groundwork. Normally, expansion of temperature will accelerate the hardening process. The anaerobic adhesives filled every surface inconsistency, resistance holes, and adequately sticks clearances upto 0.76 mm. These are capable to be connected by rapid installations in moving generation wires. So, the hardened film will have great compound protection from most liquids and gases inside a working temperature scope of –54°C to 232°C. Anaerobic auxiliary gums consolidating urethane-modified acrylic innovation were created in favor of all the demanding applications. Hence, anaerobic-based sealants in addition to string locking items were intended to restrict ordinary malleable and shear stacking. The utilization of these items frequently subjects to shear stacking. So, anaerobic gums would be able to hold out ceaseless maturing at a temperature of 232°C. Protection from salt-splash is additionally amazing.

11.4.6 Aromatic Polymer Adhesives

Significant advancement is observed in enhancing the warm and oxidative dependable qualities of natural tars at elevated temperatures. However, thermal-safe pitches and polymers were created like adhesives in the direction of addressing issues of air ship industry, rockets, and satellites, in which protection from heat moving toward 316°C is requisite for duration of the life of fortified congregations dependent on top of metals and strengthened plastic based composites. So, the oxidative solidness of natural polymers was enhanced by means of joining the fragrant and heterocyclic-based rings into the particles of polymer. Hence, the mainly significant pitches accessible in favor to utilize adhesives in elevated temperature for basic applications are (i) polyimides (PIs) and (ii) polybenzimidazoles (PBIs). These pitches were provided as prepolymers consisting of open heterocyclic rings, normally solvent, and fusible type. Finally, at raised temperatures, prepolymers experience buildup responses prompting ring conclusion and the development of insoluble in addition to infusible cured gums [8].

The temperature-based adhesives are accessible in structure of film and tape. They illustrate improved bond qualities over 260°C within air compared to epoxy-phenolic, despite the fact that the last gives better quality maintenance after presentation to water at lower temperatures. The significant burdens are their surprising expense, for the most part at least multiple times epoxy-based adhesives, whereas difficulty in dealing with or curing and the issues associated with end of volatiles amid cure so as to acquire a sans void bond. So, a long and cautious arrangement of curing ventures at continuously expanding temperature range of 316°C -371°C combined with discontinuous application and arrival of high-bracing pressure is requisite to acquire ideal outcomes. Presently, no one but PIs was utilized in temperature of about 260°C -427°C.

11.4.7 Asphalt

Asphalt is an ease thermoplastic material of profoundly temperature subordinate. The thermoplastic elastic expansion at 1–5 wt.% extraordinarily lessens the thickness reliance on heat. The expansion of thermoplastic elastic, for example, butyl elastic or polyisobutylene by 10–30 wt.% delivers really thermoplastic item with versatility, flexibility, and maximum firm quality. So, such blends are helpful as use of sealants. Black-top emulsions were utilized to build strong substance, get better water opposition, and inferior the expense on overlaying adhesives. Thus, adhesives were utilized within covering paper and supplementary pressing materials in which a water-obstruction layer was required. Therefore, one more significant usage is mainly adhesives for roofing and flooring.

11.4.8 Adhesives Based on Butyl Rubber

Rubber-based polymer utilized generally on adhesives, tackifiers, sealants, essential fasteners, and modifiers. General copolymer of isobutylene means a little measure of isoprene. So, these adhesives have moderately minimum quality and will in general display creep under burden, which are helpful in applications of bundling where their minimum penetrability to vapors, gases, and dampness were misused. It is additionally utilized as a suitable adhesive for sealant applications. However, these are commonly connected on dissolvable related arrangements.

11.4.9 Cellulose Ester Adhesives

Cellulose ester adhesives incorporate cellulose acetic acid-based derivation butyrate, cellulose nitrate, and cellulose top rate. Hence, these esters were utilized in holding wood cowhide and paper. Cellulose esters not by and large utilized with metals, non-permeable substrates, for example, cellophane (recovered cellulose) and glass are in some cases reinforced with cellulose nitrate esters connected from arrangement.

Derivation from cellulose acetic acid and their corresponding butyrate were clear as like water and extra heat safe, however lesser water safer compared to cellulose nitrate. It has preferred thermal and water obstruction over obtained acetic acid derivation in addition to wide scope of plasticizers. Therefore, cellulose nitrate is extreme and creates quality quickly, water safe, securities to numerous surfaces, and stains in daylight.

11.4.10 Adhesives Based on Cellulose Ether

It incorporates ethyl cellulose, sodium carboxy methyl cellulose hydroxy ethyl cellulose, benzyl cellulose, and methyl cellulose. Ethyl and benzyl-based cellulose were utilized as hot-soften adhesives. However, methyl-based cellulose are an extreme material; totally un-harmful, dull, and scentless build it an appropriate adhesive for nourishment bundles. So, it is fit for shaping high-consistency arrangements at low focuses as thickening operator in water-dissolvable adhesives. Therefore, hydroxy ethyl-based cellulose and sodium-based carboxy methyl cellulose likewise be utilized as thickening agent. Generally, the cellulose based ethers have reasonable for high-quality protection from high temperature. Thus, water opposition fluctuates from astounding for benzyl-based cellulose to poor for methyl based cellulose.

11.4.11 Conductive Adhesives

Fitting fillers were utilized towards delivering adhesives with maximum warm or electrical conductivity in specific applications. However, fundamental saps utilized incorporate urethanes, epoxies, polysulfones, and silicones. So, epoxies are the mainly broadly utilized resins.

11.4.12 Electrically Conductive Adhesive Materials

Electrically conductive synthetic resins were made by the development of metallic fillers and conductive carbons. Thus, carbons were utilized as finely partitioned silver flake in electrically conductive epoxies or their respective coatings. Moreover, silver has upside of having salts and oxides as conductive material, so minor oxidation and discoloring are able to endure. The resistivity procedures give much inferior esteems than strategies including slim paste lines, for example, ASTM D2739, where interfacial obstruction assumes a significant job [9]. Silver is desirable over gold as filler on the grounds that it is lesser exorbitant which has inferior resistivity. However, under states of elevated moistness and voltage, it is accounted to experience electrolytic movement towards adhesive. So, microspheres of silver-covered copper don't relocate, nor does gold. Hence, most astounding silver stacking conceivable is 85 wt.%. Carbon, i.e., graphite offers genuinely stumpy conductivities.

Similarly, other normal metallic-based fillers incorporates nickel, aluminum, and copper, which generally have specific aggravating issues. Silver is frequently utilized in flake structure building it further difficult to accomplish molecule-to-molecule contact compared with round metal based particles. So, stearate covering is connected to silver flake for improvement in dispersibility. It is likely be outgases on raised temperatures. Hence, the out-gassing might debase basic element, for example, microelectronic devices. Likewise, a few silver items remains uncoated which don't advance out items in gassing. Finally, structure of copper and aluminum oxide-based films decreases electrical conductivity via hampering molecule-to-molecule contact [10]. These are always utilized in microelectronic congregations [11]. These applications incorporate holding parts to plated through gaps on printed circuits, connection to fine lead wires to printed circuits, metallization on fired substrates, establishing metal undercarriage, electroplating puts together, holding wire prompts header pins, wave-manage tuning, and gap fixing. These adhesives were connected like replacements in spot welding and their corresponding temperatures to develop overabundance opposition at the weld in light of oxide arrangement. Likewise, an additional application is in ferroelectric gadgets utilize to attach anode terminals to the precious stones in stacks. So, these adhesives supplant binds and welds in which precious stones will in general be kept by patching and welding temperatures.

11.4.13 Thermally Conductive Adhesives

In these applications, temperature ascends because of development of heat from cylinders, resistors, transformers, and so forth, in high-thickness circuits are regularly basic and a reason for concern. Plan contemplations must incorporate thermally conductive divisions for expelling thermal heat from the hardware included. Inside the confined hardware, printed-circuit board and non-typified thermal sinks reinforced set up is a basic arrangement. For this situation, aluminum was normally favored as thermal-sink material as a result of less weight and maximum warm conductivity. In the event that great dielectric characteristics are necessary, an elevated centralization of inorganic fillers is able to utilize. Thermal conductive-based epoxy framework utilized as part of adhesive, just like dissimilar purposes have warm conductivity and volume resistivity. However, fillers incorporate silica, alumina (aluminum oxide), boron nitride, beryllia (beryllium oxide), and other unspecified inorganic oxides. Thus, boron nitride, a fantastic decision as thermally conductive inorganic filler with the exception of that its substance achieves a most extreme of 40 wt.% mainly in epoxy saps. The resulting items were dependably thixotropic glues. Hence, beryllia powder having brilliant warm conductivity independent blended with a sap cover to drops its conductivity radically. So, it is likewise exceedingly poisonous and lifted in expense.

11.5 Resin

11.5.1 Unsaturated Polyester Resin

In general, this includes chain linked polyester encasing twofold securities and a vinyl assisted monomer. Copolymerization results revealed the presence of vinyl monomer along with twofold reinforced polyester. Unsaturated polyester resins have a place with the gathering thermosets. Unsaturated polyesters are created from the buildup bowl, which are weak at surrounding temperature and hard to deal with. Polyester is fabricated from a plant and blended with the vinyl monomer of fluid state which are generally thick at encompassing temperature and contain 60% of styrene substance. Therefore, a blend comprising of unsaturated polyester and vinyl polymer named as unsaturated polyester resin [12]. The parameters for copolymer synthesis and resin qualities like swelling and obvious thickness for various polymers are recorded in Tables 11.1 and 11.2 [13].

11.5.2 Monomers

In light of its compositions, cluster of monomers are existing in two modules, one as polyester segments and other as vinyl monomer segments.

11.5.2.1 Unsaturated Polyester

Soaked alcohols like trimethylolpropane, neopentyl, glycerol, propylene, diethylene, tetrabromobisphenol A, trifunctional alcohol, trimethylolpropane mono alkyl ether, ethylene glycol, trifunctional alcohol, and immersed anhydrides/acids that include tetrabromophthalic anhydride, terephthalic acid, chlorendic acid, tetrabromophthalic anhydride, ohthalic anhydride, and isophthalic acid are commonly utilized as monomers for delivering unsaturated polyester resins [14].

11.5.2.2 Alcohol Constituents

1,2-propylene glycol along with ethylene glycol is the regular alcohol ingredients. Air drying properties of ether including alcohols are relatively superior to other elements. Polyesters dependent on diols has been delivered via the transesterification reaction of diethyl adipate by means of unsaturated diols like 4-diol, 2-butyne-1, 4-diol, and cis-2-butene-1. This process possesses most appropriate technique of developing polyesters in unsaturated mode as opposed to the direct poly-condensation reaction. The most existing aliphatic unsaturated diol like 4-diol, 2-butyne-1, 4-diol, and cis-2-butene-1 has been utilized to deliver a portion of the significant polymers for medicinal and other related applications.

11.5.2.3 Constituents Like Anhydride and Acid

Wide-ranging mechanical unsaturated polyester normally developed through phthalic anhydride, 1,2-propylene glycol, and maleic anhydride. So, styrene is mainly utilized as vinyl assisted monomer. The anhydride

Table 11.1 Parameters of copolymer synthesis [13]. (With kind permission obtained from RightsLink, Elsevier License).

Polymerization	Monomer mole ratio (MA/ST/DVB) (%)	Diluent Hep/Tol (%)	Transfer agent type/(%)[a]	Suspension agent type/(%)	Yield (%)
1.	0/80/20	70/30	Triton™/20	Gelatin/0.5 and HEC/0.5	95
2.	30/50/20	70/30	TBA/20	Gelatin/0.5 and HEC/0.5	60
3.	60/20/20	70/30	Triton™/10	Gelatin/0.2	65
4.	30/50/20	100/0	Triton™/30	–[b]	75
5.	30/50/20	100/0	Triton™/20	–[b]	82
6.	30/50/20	100/0	Triton™/10	–[b]	72
7.	30/50/20	100/0	TBA/30	–[b]	64
8.	30/50/20	100/0	TBA/20	–[b]	73
9.	30/50/20	100/0	TBA/10	Gelatin/0.4	57

(Continued)

Table 11.1 Parameters of copolymer synthesis [13]. (With kind permission obtained from RightsLink, Elsevier License). (*Continued*)

Polymerization	Monomer mole ratio (MA/ST/DVB) (%)	Diluent Hep/Tol (%)	Transfer agent type/(%)[a]	Suspension agent type/(%)	Yield (%)
10.	30/50/20	100/0	Triton™/10	Gelatin/1.0	75
11.	30/50/20	100/0	TBA/10	Gelatin/1.0	73
12.	30/50/20	100/0	Triton™/10	Gelatin/0.5 and HEC/0.5	58
13.	30/50/20	50/50	Triton™/10	Gelatin/0.5 and HEC/0.5	70
14.	30/50/20	100/0	Triton™/10	Gelatin/0.7 and HEC/0.7	70

Hep - heptane; Tol - toluene; TBA - tetrabutylammonium perchlorate; TritonX-100TM - [polyoxyethylene-(10)-isooctylphenyl-ether]; HEC - 2-hydroxy-ethyl-cellulose; MA - methacrylamide; ST - styrene; DVB - divinylbenzene.
Polymerization conditions: time: 24 h; temperature: 60°C; initiator: AIBN (1% M); monomer molar ratio: 30% methacrylamide/50% styrene/20% DVB; dilution: 100%; Na_3PO_4: 2% wt./v; organic phase/aqueous phase relation: 1/4.
[a]Molar % in relation to methacrylamide, [b]No suspension agent.

Table 11.2 Swelling and density characteristics of unsaturated polyester resin [13]. (With kind permission obtained from RightsLink, Elsevier License).

		Swelling (%)		
Polymer*	**Apparent density (g/ml)**	**Toluene**	**Methanol**	**Water**
1.	0.46	80	10	10
2.	0.45	85	10	10
3.	0.25	60	55	50
4.	0.26	28	40	25
5.	0.29	25	35	20
6.	0.33	35	35	20
7.	0.31	35	28	15
8.	0.30	38	13	13
9.	0.39	55	25	15
10.	0.29	43	15	15
11.	0.36	40	20	15
12.	0.44	43	30	20
13.	0.63	115	30	20
14.	0.34	33	37	27

*The polymers are the same described in Table 11.1.

exclusive of phthalic anhydride delivers polyester with thickness of twofold attachment along with cluster of polyester chains. The unsaturated corrosive segment is constantly weakened with acid and non-polymerizable twofold securities. Generally, aromatic twofold bonds likewise won't polymerize with vinyl segments. Fumaric corrosive copolymerizes with styrene, yet fumaric corrosive is costlier compared to maleic anhydride. An additional significant perspective focused that condensation duration of those acids expelled water. Anhydrides, which are favored for acids resulted in superior reactivity, whereas acids can't structure like an anhydride. These mixes don't consolidate as quickly as anhydrides. Resting on the opposite side, polyesters obtained from those acids are progressively steady contrasted with segments produced using phthalic anhydride and also utilized within forceful situations, top paint, and gel coats. At that point, the glass fiber fortified overlays are connected sequentially and the top coat is at long last painted [15].

11.5.3 Vinyl Monomers of Unsaturated Polyester Resins

The vinyl monomer fills in as a decent dissolvable solution for the polyester and decreases its thickness. It is likewise an operator of copolymerization throughout restoring. The vinyl assisted monomers for polyester resins like methyl acrylate, triallylcyanurate, p-vinyl toluene, etc., present in unsaturated resins are mainly utilized alongside polyester monomer to deliver polyester based resin.

11.5.4 Styrenes

This is the mainly utilized vinyl assisted monomer designed for creating unsaturated polyesters. It contains oncogenic potential, so substituting styrene via several additional monomers for a long time. Increment in styrene expanded the unbending nature of the material. Whereas, α-methyl styrene shapes less responsive radicals and in this manner hinders the relieving response. Hence, α-methyl styrene is best appropriate for lessening the pinnacle temperature during the curing process. Characteristics of unsaturated fat and styrene related vinyl monomers at normal room temperature and post-relieved temperature are recorded in Table 11.3 [16]. Polar vinyl monomers like vinyl pyridine improved the attachment of the polyester to glass strands, which is helpful in deciding delamination in composites.

11.5.5 Acrylates and Methacrylates

Acrylate improves outside strength, whereas methyl methacrylate improves the optical properties. Also, authors considered the polymerization of methacrylate by utilizing light as a source. The refractive indicator is capable of altered through styrene and methyl methacrylate blends which is closely related to glass [17].

11.5.6 Vinyl Ethers

Distinctive vinyl ethers mainly utilized like exchanges in favor of styrene. The unsaturated polyesters utilized broadly in emission treatable pieces and especially in coatings [18]. Exceptional plans containing zero styrene

Table 11.3 Characteristics of vinyl esters based on styrene and fatty acid cured at room temperature and post-cured with a temperature from room temperature to 200°C at 5°C/min [16]. (With kind permission obtained from RightsLink, Elsevier License).

	35% Co-monomer			45% Co-monomer		
Co-monomer	**Tg (°C)**	**E' at 35°C (MPa)**	**Mc (g/mol)**	**Tg (°C)**	**E' at 35°C (MPa)**	**Mc (g/mol)**
Styrene	147	2.9	328	142	2.6	642
MLAU	79	2.0	211	71	1.5	304
MOA	76	1.6	240	75	1.1	373
MLIN	72	1.6	290	69	1.0	400

and existing triethylene glycol di-vinyl ether be able to utilize in support of pertaining gel coats. Mainly, propenyl ethers earn a lot simpler towards getting ready contrasted with gathering of vinyl ethers. However, these are created through isomerization reaction of allyl ethers. Because of antagonistic impact of methyl bunches in propenyl ether atoms, they get anticipated considerably fewer responsive than the vinyl ether analogs.

11.5.7 Fillers

The couple of instances of fillers incorporate powder of calcium carbonate, alumina, silica, glass, and earth minerals, silica sand talc, white marble, marble scraps, glass dabs, mica, yarn, silica sand, cellulose, waterway, and squashed stone. On account of some of the fillers, translucency is revealed during curing process. Normal fillers are wood flour dirt, bentonite, elastic particles, calcium carbonate, glass dabs, fly fiery remains, and nanocomposites. Generally, fillers decline the expense and interchange various mechanical characteristics of the relieved resources.

Bentonite and ca-based bentonite is utilized during production of composites based on unsaturated polyesters. Expanding the quantity of filler in reins by a consistent proportion of styrene and polyester improved their mechanical characteristics. Most extreme estimations of hardness, compressive quality and heat conductivity are nearly around 22.7% for styrene [19].

11.6 Polyurethanes

An isocyanate segment and a diol segment are the two major parts of polyurethanes. The diol segment may comprise of a polyether or polyester end topped diol. Their corresponding structure of urethane might exist as esters of acids. Their development is finished by means of the expansion in amines and organometallic composites. Polyurethanes additionally discover use in numerous therapeutic applications. They are also utilized as adhesives and as coatings to an enormous degree.

11.6.1 Monomers

Monomers associated to create polyurethanes comprise two kinds, namely, (i) diisocyanates and (ii) polyols.

11.6.1.1 Diisocyanates

The preparation process begins by means of an amine, phosgene, and aliphatic. Similarly, isocyanate will be generated as a result of the expulsion of hydrochloric (HCl) acid.

11.6.1.2 Phosgene Route

The amalgamation course by means of phosgene was created in 1884 by Hentschel, despite the fact that Wurtz in 1848 uncovered isocyanates. The synthesis keeps running by two essential advances, namely,

1. Arrangement of carbamic chloride
2. Removal of HCl

11.6.1.3 Phosgene-Free Route

Synthesis procedure of phosgene free process was created in view of the risks in taking care of phosgene. The combination begins with nitrobenzene and consecutively the ethyl urethane straightforwardly delivered with carbon monoxide (CO) and ethanol. However, the presence of urethane dimerized by means of a carbonylation response. Through temperature expansion, the urethane decayed onto isocyanate and alcohol. The incredibly unpredictable isocyanates are increasingly poisonous. At the time of hardening, there is additionally an arrival of the unreacted isocyanate. The discharge likewise relies upon the isocyanate reactivity, which is distinguished within blends. On account of very high reactivity along with dampness, investigation needs uncommon systems [20].

11.6.1.4 Polyols

The second fundamental monomer next to diisocyanates is polyols, which is of two types as follows:

1. *Polyether polyols:* This is one of the most generally utilized monomers. Utilized monomers are ethylene oxide, butylene oxide, propylene oxide, tetrahydrofuran, and propylene oxide [21].
2. *Grafted polyols:* Interchanging styrene or acrylonitrile into polyphenylene oxide (PPO) results in developing copolymer

polyols. At first, unadulterated acrylo-nitrile be utilized in favor of uniting, yet so created copolymer polyols causes staining issues in adaptable foams. Consequently, styrene/acrylonitrile copolymer polyols were created [22].

11.6.1.5 Vinyl Functionalized Polyols

Vinyl moiety is an additional significant technique to functionalize the polyols, which accomplished by response of the polyols with maleic anhydride. Clearly, the usefulness of the polyols must be more noteworthy than the hydroxyl gathering, since hydroxyl gatherings are evaporated. On the off chance that the vinyl functionalized polyols polymerizing by including vinyl monomer blend, the swinging vinyl bunch polyols have an influence in the polymerization process. Compare with the vinyl assisted polymer, brush resembling structure is developed and teeth of "brush" creature the polyols. Styrene is generally hydrophobic and in the sight of cutting edge transformation, foundation of brush might breakdown to capitulate a round shape.

11.6.1.6 Polyols Based on Modified Polyurea

Polyols based in urea and adjusted polyurca arc a suitable alternate, which normally blended as a two phase reaction. In the initial period, amine-based alcohol is permitted towards responding by means of a surplus of diisocyanate to develop urethane assembly. Overabundance of isocyanate creates arrangement of an end-topped prepolymer. On account of this, they obtained as solely urea isocyanates bunch in spines. Similarly, in amino based alcoholic isocyanates, the observance of urea and urethane are framed as spines. In general, appropriate diamines are hydrazine, ethylene diamine, and so forth. During next stage, polyols in molar overabundance as for the unreacted isocyanate gatherings was included. The pending isocyanate gatherings respond with the hydroxy gatherings to frame chain-broadened polymeric polyols. Obtained response with isocyanates continues quicker and contrast with response of polyols [23].

11.6.1.7 Polyols Based on Polyester

Monomeric blends for this polyols are glycerol, 1,2-propylene glycol, adipic and phthalic acid, diethylene glycol, oleic acid, 1,1,1-trimethylolpropane, and 1,1,1-trimethylol propane adipic acid.

11.6.1.8 Acid and Alcohols-Based Polyesters

Generally, these are created by preheating diol at 90°C followed by including harmful acids onto it, which finish the process at the elevated heat range upto 200°C. The water was removed after the process by passing inert gas or vacuum. The buildup is a harmony process and Schulz-Flory appropriations of atomic weights are acquired. However, chemical buildup process is carried out through acids, bases along with progress metal mixes. The increment in hydroxyl numbers of adaptable to inflexible foams from 60 to 400 mg of potassium hydroxide per gram (KOH/g). Acids for delicate foams are aliphatic based which expands the unbending nature.

Isophthalic acids are utilized in hard coatings and bindings. Such foams demonstrated upgrade in fire safe properties. However, polyesters are fabricated by transesterification reaction of dimethyl terephthalate. Likewise, polyethylene terephthalate (PET) dissipates resources like polyester-based filaments and or soda containers which are reused by glycolysis process to acquire proper polyols.

11.6.2 Rectorite Nanocomposites

Dirt mineral, rectorite, is an ordinary interstratification reaction between dioctahedral mica and a dioctahedralsmectite. It is mainly utilized to obtain intercalated thermoplastic polyurethane elastic nanocomposites through melt processing method. The researchers obviously elucidate those composites comprising lesser measures of earth minerals are mainly intercalated or part shedded by utilizing X-ray beam diffraction and transmission electron microscopy (TEM). Finally, the result shows upgraded mechanical characteristics of the composites [24].

11.6.3 Zeolite

These are utilized to design for adjusting the polyurethane layer structures to advance their corresponding characteristics. Layers with these substances somewhere in the range of 10% to 70% have been readied. The planning technique initiates anisotropy in the films. The layers have a hilter kilter structure comprising of the top skin, which incorporates the dynamic layer, the substructure, and the base skin.

11.7 Epoxy Resins

These are mainly developed from oligomer having atleast two parts of epoxide gatherings and a single curing operator, more often than not contain any amine compound. Numerous monographs are accessible in the market based on epoxy resins.

11.7.1 Monomers

11.7.1.1 Epoxides

Epichlorohydrin, a monomer, is utilized in favor of developing glycidyl ethers and esters. This suitable monomer is produced through propene by means of allyl chloride. However, a generally available epoxide portions are poly (butadiene) epoxides, epoxide, glycerol diglycidyl ether, epichlorohydrin, glycerol diglycidyl ether, butadiene diepoxide, and 1,4-butanediol diglycidyl ether. The relieving of cycloaliphatic epoxides continues effectively with anhydrides however is excessively moderate with amines.

11.7.1.2 Hyper Branched Polymers

Hyper branched polymers are exceedingly spreaded macromolecules which can set up by means of a solitary advance polymerization reaction. Numerous polymers of the same kind are otherwise called dendrimers. Additionally, star-like or brush structured polymers have a place with the class of hyper branched polymers. However, hyper branched polymers are developed from dendritic, direct, and terminal units [25]. They can be combined by means of three courses:

1. Step-development poly condensation process of ABx based monomers
2. Self-gathering vinyl assisted polymerization reaction on ABx based monomers
3. Multi-spreading ring-opening process for idle ABx based monomers

General techniques for blend manufacturing permit a broad range in types of polymer. Extraordinary properties can be bestowed through reasonable end topping responses. These kinds of polymers have special characteristics for trademark for dendritic macromolecules like low down consistency, great dissolvability, and usefulness. These are utilized in restorative fields because of transporters in natural mixes. Hyper extended polymers are simpler to combine in huge amounts and are utilized as anti-plasticizers, tougheners, curing reagents, and plasticizers. Hyper spread polymers (HBP) along with hydroxyl gatherings are able to start relieving via a proton contributor complex. While relieving tetra utilitarian epoxy resins, actuation vitality is lesser than epoxy framework in the midst of direct polymers. HBP firmly upgrade the relieving rate because of the reactant impact of hydroxy gatherings.

The prolonged gel time with expanding usefulness commencing diglycidyl ether of bisphenol A (DGEBA) to tetraglycidyl-4,4′-diaminodiphenylmethane (TGDDM). Similarly, hydroxyl-functionalized HBP decreased the gelling duration of mixes on account of quickening impact of hydroxyl gatherings toward the epoxy relieving response. Star-like structured epoxy polymers were developed by means of polymers with epoxy monomer. The expansion of little measures of HBP towards an epoxy framework upgraded their durability. However, the basic damage vitality discharge rate of DGEBA resin is capable of expanded as a result of 6 factors by including 5% HBP. At higher focuses, a stage division is shown by dual glass change temperature ranges. Inside composite materials, HBP changed resins permit superior volume divisions of filaments for delivering void-free overlays during contrast with unmodified resins.

11.7.2 Epoxide Resins Based on Liquid Crystalline Structure

At first, couples of specialized conditions relating to fluid gems were reviewed. Fluid precious stones were found by Austrian physicist and botanist Friedrich Reinitzer, who already found that cholesterol benzoate, will not dissolve onto a reasonable fluid. On advance heat, turbid fluid twisted all of a sudden clear. So, this change point is known as the clearing point. Notwithstanding the basic conditions of conglomeration, the fluid crystalline state was set up.

11.7.3 Liquid Crystal

Rod-like molecules are used for generating liquid crystals (LC), which is generally dealt with mesomorphic phases. In which the materials with the intention of framing those stages are called mesogens. Isotropic, a standard liquid, has its properties for free of bearing. Similarly, fluid gem be orientated like an anisotropic fluid, which implies that atoms were situated ideally during a specific course. Such an anisotropic liquid be a pneumatic fluid gem. The particles set up inside the layers however the atom does not have any fixed positions. Fluid crystalline polymers show various improved properties in correlation with customary plastics and expanded the flexible moduli at high temperatures, detoriates beginning temperatures, diminished the coefficients of thermal development and dissolvable ingestion. Appropriate epoxide monomer mainly depends on biphenyl moietics. However, Micro-Brownian movement of polymer chain improved by expanding the convergence of mesogens. This impact causes an increment in the warm disintegration beginning temperature ranges, reducing in coefficient of warm development and abatement on water ingestion. During the curing process, cross-connected systems were achieved through fluid crystalline properties.

Additionally sulfanilamide and diglycidyl ether blend of bisphenol A do not demonstrate LC characteristics [26]. An isotropic fluid is shaped over the softening point. Exactly, during the process among amine and epoxy continues, this surfaces were created to secure the cross connected system through the nematic course of action.

11.7.4 Liquid-Based Rubbers

The expansion of rubbers onto epoxy adhesives improved the crack obstruction, attachment to slick surfaces, flexibility, and strip quality. Fluid rubbers are utilized as some of the toughening operators. Fluid elastic modifiers were fundamentally miscible with the epoxy. At curing stage, a partition happens in the process. Carboxy-ended butadiene/acrylonitrile copolymers (CTBN) are dominatingly reasonable as a result of miscibility within a few epoxy resins. On the off chance that a CTBN isn't pre responded, the carboxylic acid gatherings be able to respond at hardening stage. Nitrile rubber (NBR) specifically with maximum acrylonitrile substance is additionally fit for tougheners; however, adding high quantity in the rubber gives great similarity among NBR and the epoxy resin material [27].

11.7.5 Silicone-Based Elastomers

Silicone-based elastomers dropped the ideal mechanical characteristics in the elevated temperature and low temperature locale. These are prevalent in their qualities. So, these rubbers are totally immiscible by means of epoxy resins and can't exist thereof. The expansion of silicone added poly(methyl methacrylate) (PMMA) is compelling towards balance out the boundary of silicone elastic and epoxy resin which assist to break down silicone-based rubber in epoxide complexes. The sub-atomic load of the silicone fragment influences the adequacy of the compatibilizer. While expanding molecule measurement of the silicone, the break durability diminishes and dips under the unmodified gum.

11.7.6 Rubbery Epoxy Compounds

Rather than elastic fluid, rubbery-based epoxy particles developed from aliphatic epoxy resin are mixed by means of another epoxy gum towards toughening operators. The impediments of epoxy-CTBN have maximum consistency. In any case, there are likewise low-thickness types accessible. In car, aeronautics and the related enterprises, this utilizes metallic materials along with polymeric-based materials. In some parts of such applications, great tribological characteristics are essential. The incorporation of fluorinated poly(aryl ether ketone) (F-PEK) improved the same. This kind of fluorinated polymers are always recognized as low grating resources, which generally emerges because of their stumpy surface energies.

11.7.7 Adhesion Improvers

Hybrid resins of epoxy polyurethane are utilized for adhesives of high quality. Also, elastomeric altered resin solutions are utilized as glues.

11.7.8 Unsaturated Polyesters

In epoxy blend, curing by means of differential filtering calorimetry showed a superior rate consistent compared to unadulterated epoxide materials. Also, it is trusted that the hydroxyl assembly in the mix gives an ideal synergist condition to the epoxide relieving. The understanding of the consistency improvement recommends that an interlock between the two developing systems exists to expand the thickness. The presentation of unsaturated polyester into epoxy gum improved the strength however decreased the glass transition temperature.

11.7.9 Functional Peroxides

The anhydride's condensation reaction along with polyethylene glycol (PEG) and tertbutyl hydroperoxide developed peroxy ester oligomers as a result [28]. The subsequent esters include carboxylic and peroxy groups, which can mix and utilize like curing operators [29].

11.7.10 Acrylics

The polymer systems with profoundly polymerizable segment act as appropriate restoring specialists. The temperature of around 65°C and 80°C are mainly accomplished as curing range. However, the energy for restoring epoxide happens due to a blend with uncatalyzed bimolecular and catalyzed atomic response. Within the blend, lower rate constants were observed compared to different frameworks, whereas the higher actuation energies were also depicted. The chain entanglements sandwiched between two systems root a steric obstacle on behalf of the relieving procedure. Finally, the vitrification process limits chain portability which is reflected to diminish rate constants.

11.7.11 Bismaleimide Based on Modified Urethane

This is acquainted plus joined halfway with epoxy oligomers through polyurethane blending operators. Meanwhile, a mass polymerization strategy can be utilized to get ready interpenetrating systems [30]. The rigidity expanded to a most extreme incentive with expanding the polymer content, which diminishes with further expanding the same. Infrared spectral investigation was utilized to affirm the blending of polyurethane into the epoxy skeleton [31]. The created frameworks were described by heat, morphological, and mechanical examinations.

11.7.12 Hybrid Materials Comprising Organic and Inorganic Compounds

Natural inorganic mixture interpenetrating systems were created from an epoxide-based amine framework and tetraethoxysilane (TEOS). However, the energy of silica structure development on the natural grid and its last structure and morphology rely upon the strategy for generation of the interpenetrating system. During sol gel, hydrolysis and polymerization reaction of TEOS were performed at nominal room temperature in isopropyl liquor. The crossover systems know how to set up by two strategies. The strategies for arrangement of hybrid system were also examined [32].

In the one-advance methodology, all response parts are changed all the while. During two-advance strategy, it is hydrolyzed within the initial step, blended through the natural epoxy segments and polymerized beneath the development of silica/epoxide systems. Huge smaller silica totals around 100 to 300 nm widths were shaped through the one-arrange procedure of polymerization reaction. Also, TEOS fractional hydrolysis impacts a speeding up of the gelation. The expansions in modulus through two requests of extent were accomplished at 10% of silica content.

11.7.13 Poly (Ether Ether Ketone)

Poly Ether Ether Ketone (PEEK) is an intense, superior, semi-crystalline thermoplastic polymer along with great heat and mechanical characteristics. Due to crystalline nature, the polymer is hard to mix with epoxy resins. A few strategies consisting of dynamic mechanical analysis (DMA), Fourier transform infra-red (FTIR) spectroscopy, and scanning electron microscopy (SEM) show that the relieved mixes are homogeneous [33]. With expanding PEEK-C content, the ductile properties and break toughness factor of the mixes diminished somewhat.

11.7.14 Epoxy Systems Comprising of Vinyl-Based Polymers

Other than unadulterated frameworks, blended frameworks, for example, epoxy acrylates, are being used, which can cure with radical photo initiators.

11.7.15 Characteristics

Mechanical abilities of this resin are related to follow back the formation of monomers. It mainly relies upon adaptability and the crosslinking thickness. Normally, epoxy resins recoil at the season of restoring not as much as vinyl resins. Although, it is critical toward separating the shrinkage happens ahead of gelling and in the wake of gelling. Researchers examined temperature coefficient of polymeric materials and found that impact of curing process. So, elevated temperature impacts the curing process further [34].

These resins normally have very low viscosity, even after curing; it displays high transition temperature, low elongation at break and toughness due to its high crosslinking density. Epoxy resins depict great electrical characteristics, which normally influence by means of the moisture content. Similarly, addition of metal particles, namely, silver and copper, makes the resins electrically conductive for developing strong bonds within the surfaces and owing its usage mainly in adhesives. Normally, this also has exceptional opposition to organic and inorganic solvents, acids, salts, bases, and chemical agents.

11.7.16 Hybrid- and Mixed-Bases Polymers

Normally, hybrid and or blended polymers incorporate epoxy based silicone, maleimide, and urethane hybrid polymers.

11.7.17 Copolymers-Based on Epoxy-Siloxane

Epoxy-siloxane polymer blended by means of diglycidyl bisphenol A ether and jointed tightly will expand the portability of the cross connected system and the thermal solidness [35]. Joining the copolymers by pendant epoxide rings mainly incorporates the hydrosilylation reaction of poly (methylhydrosiloxane) through allyl glycidyl ether. Amino propyl-ended polymer mixed within an epoxy resin demonstrates extraordinary oil as well as water repellency properties in coatings. Further, the strip quality of weight delicate cement appended in the direction of the altered epoxy based resin additionally diminished. The obtained copolymers included measures of 5 ca. phr, proficiently decrease the stationary grinding coefficient of cured mixes.

11.8 Phenol Formaldehyde Resin

The most seasoned thermosetting polymers are phenolic resins. Regardless, they have numerous mechanical applications in parts, for example, automobiles, computing, aviation, and construction. Generally, these resins are prepared using condensation process of a phenol, in which water is created as a result. Ordinarily, phenol and formaldehyde have been utilized to create phenolic resins with explicit characteristics, for example, adaptability and reactivity. Cardanol formaldehyde resin is broadly utilized alongside phenol and cardanol. The physico-chemical characteristics of the same are also classified in Table 11.4 [36]. The various assortments of phenolic resins existing are very huge because of the proportion of phenol and aldehyde; however, response temperature and catalyst selection procedure can be differed. The two broad classifications of phenolic resins are (i) novalak and (ii) resol. A resol resin, which is of single stage, is regularly delivered with overabundance of a phenol, formaldehyde, and alkaline compound.

Table 11.4 Physico-chemical properties of cardanol and cardanol formaldehyde resin [36]. (With kind permission obtained from RightsLink, Springer Nature License).

Properties	Cardanol	Cardanol formaldehyde resin
Color	Dark brown	Dark brown
Odour	Phenolic	Phenolic
Specific gravity (g/cc at 30°C)	0.9285	0.9333
Viscosity (intrinsic)	3.13×10^{-2}	3.28×10^{-2}
Hydroxyl value	181.5	157.2
Iodine value	223.2	221.8
Molecular weight	302	1550
Number of hydroxyl groups	1	5
Thin layer chromatography: R_f values		
(a) Petroleum ether: diethyl ether (7:3)	0.82	0.78
(b) 100% benzene	0.77	0.75
(c) Benzene: chloroform (1:1)	0.81	0.76

The response is proscribed en route for making a non-crosslinked resin with the purpose of relieving the heat lacking extra catalysts to shape a 3-dimensional cross-linked insoluble, infusible polymer. Interestingly, novolak-based resins created by formaldehyde incorporate the solid mineral acids like phosphoric corrosive, sulfuric corrosive, and hydrochloric corrosive alongside natural corrosive catalysts, for example, oxalic corrosive, p-toluene sulfonic acid, and inorganic salts. It creates a thermoplastic polymer which can liquefy yet won't crosslink lead the utilization of thermal treatment alone. Subsequent novolak-based thermoplastic resin be crosslinked through the expansion of novolak relieving specialist. There are different kinds of novolak-based resins through various ortho-to-para proportions of methylene relationship like high ortho and high para novolak based resins (HON and HPN) and general purpose novolak based resins (GPN) [37]. No extra curing operators will be required for resol resins, which can only healed by heat receptive, and it has a very low timeframe of realistic usability.

11.8.1 Monomers

Bisphenol A, cresols, xylenols, and bisphenol B are the appropriate phenol subsidiaries for utilization in phenol formaldehyde resins. Likewise, para-formaldehyde, acetaldehyde, propionaldehyde, butyraldehyde, furfural, furfurol, and glyoxaltrioxane are the suitable formaldehyde derivatives for phenol formaldehyde resins.

11.8.2 Phenol

The cumene peroxidation is the favored course to phenol, representing over 3/4th of world creation. Additionally, cumox process generally comprises of

1. Fluid stage cumene oxidation to cumenehydroperoxide (CHP)
2. Disintegration of the obtained product to phenol and acetone

Phenolic resins, caprolactam, and bisphenol A are mainly utilized in production of phenol. It is likewise utilized in the assembling of numerous items comprise protection equipment, adhesives, elastic, polishes, ink, toys colors, scents, cleansers, paint, and enlightening gases.

11.8.3 O-Cresol

O-Cresol is utilized generally in the middle for the generation of epoxy resins, pharmaceuticals colors and pesticides, yet in addition as a part of cleaning specialists and disinfectants. It is promptly recyclable and has a little bioaccumulation. Around, O-Cresol of 60% is gotten from coal-tar and unrefined petroleum by utilizing established systems, for example, stripping, refining, and fluid extraction. The staying 40% is gotten artificially through the alkylation reaction of phenol with methanol.

11.8.4 Formaldehyde

It is a most fundamental modern industrial chemical utilized for the planning of an assortment of many synthetic chemicals. Normally, it is color-less and very combustible gas for commercial applications. Formaldehyde is utilized for the most part inside the amalgamation of resins along with phenolic-formaldehyde, urea formaldehyde, pentaerythritol, and different resins. However, only 6% employments be identified with compost generation. Formaldehyde discover application in an assortment of businesses, including the cleanser, sustenance, restorative, elastic, metal, wood, medicinal, cowhide, oil, manure, and rural ventures.

11.8.5 Multihydroxymethylketone

Multihydroxymethylketone be the reaction results of ketones among a huge abundance of formaldehyde, which are utilized as reactive solvents and other different applications. Phenol blend in a multihydroxymethylketone generates an uncommon sort of an altered phenol/formaldehyde resin.

11.8.6 Basic Resin Types

11.8.6.1 Novolak Resins

A novolak resin comprising of atleast one phenol subordinates in addition to aldehyde. These are utilized during elastic arrangements to provide the creation of belts, cylinders, and tires, which be capable of fortify the elastic arrangements. The reinforcement is clarified by the development of a three-dimensional system inside the elastic components after relieving them [38].

11.8.6.2 Resol Resins

Generally, these are regularly ended up through buildup polymerization within the sight of a medium at various temperatures ranges between 40°C and 100°C. Thus, alkaline medium is fundamental, a corrosive component would be utilized and an uncontrolled relieving during the readiness of the prepolymer would happen. Then again, restoring of the resol prepolymer possibly will be accomplished by fermentation. Because of little yield containing free monomers, buildup under the ordinary response conditions holds a high level of free monomers. These are unpredictable and profoundly very harmful. Diminishing the dimension of free monomers in such resins, in this manner decreasing their outflows into the earth-based application procedures has been a standout among the most vigorously investigated regions by both phenolic resin makers for a long time. Resol alludes to phenolic resins that hold valuable reactivity rather than the relieved resins. At this juncture, the item is completely dissolvable and fusible at less than 150°C in at least one basic solvent, for example, alcohols and ketones [38].

11.8.7 Fillers

Numerous kinds of fillers are referred to for phenol-based formaldehyde (PF) resins, for example, natural fiber materials, glass strands, and ceramic material.

11.8.8 Reinforcement Based on Jute Fibers

This material is capable of reused into compound utilizing PF resin of 12% to 30%. So, the dimensional dependability of created composites was enhanced by acetylating or through steam treatment of jute material. Thus, the jute material is better than acetylation in enhancing the dimensional steadiness. This material displays considerably fewer irreversible and reversible enlargements compared to untreated jute material.

11.8.9 Applications and Uses

These are utilized on the way to build an assortment of items consist of desired wood items, for example, pressed wood, designed timber, hard board, fiber board, and arranged strand board. Different items incorporate covers, fiberglass protection, grating coatings, foundry covers, oil recuperation froths, and grinding fasteners. They are utilized at the same time as paper soaking resins in favor of overlay, oil channels, paint roller tubes, and so on. These resins utilized to compose open or closed foams, which are basically utilized to construct botanical foam, bolsters used for holding bloom stems in fluid water. These foams are able to drench up water ordinarily and estimated its unique load to give water to the blossoms. These froths are principally open cell. Different uses for phenol formaldehyde froths are thick froths to hold adornments in addition to build molds in favor of foot prosthetics. Shut cell froths discover applications in boundary and protection materials. Additional employments of resins incorporate grinding covers, grating folios, and phenol formaldehyde covered foundry sand covers [39].

11.8.10 Binders for Glass Fibers

Bulk fiber for protection and partnered purpose and continuous fiber or strengthening strands are the two major types for developing glass filaments. In either structure, the crude fiberglass is rough and delicate. Harm to the single glass strands can happen because of the self-grating movement of lone fiber disregarding or communicating by means of an additional fiber. However, the subsequent exterior deformities root to decrease in the general mechanical properties of fiberglass. So, these were created to avert the above issues. A trademark folio may avert the damaging impacts of self-scraped area without repressing the general adaptability of the completed glass fiber item. Great opposition and versatility to extraordinary states of raised stickiness and heat are gainful in perspective on the wide assortment of utilizations of glass fiber fastener arrangements. The measure of folio available in a fiberglass item is reliant on a few elements together with the sort of administration required, the item shape, foreseen natural factors like temperature and compressive quality prerequisites.

11.8.11 Phenolic Binders

Customarily, the exhibition constraint essential for protection filaments were fulfilled uniquely by phenol formaldehyde resins. In this manner, glass fiber fasteners were solely founded on top of phenol formaldehyde resins. It commonly incorporates aminoplast resins like silicone mixes, solvent oils, melamine and urea, stabilizers, and wetting specialists. Commonly, phenolic folios contain a lot of low atomic weight group together with phenol, formaldehyde, and combination of both, i.e., phenol formaldehyde.

During curing, unstable less sub-atomic weight segments are discharged onto the air in considerable capacity as unpredictable natural mixes, i.e., volatile organic compounds (VOCs). The way toward assembling fiberglass normally includes showering huge amount of PF fasteners into elevated capacity of air streams and restoring the item in convection broilers that include huge quantity of air. However, fiberglass makers enclose a critical call for diminishing their VOC emanations as to formaldehyde. Endeavors to change over free formaldehyde into less disagreeable and hazardous synthetics have included the expansion of urea or smelling salts. These increments were expected to change over free formaldehyde to hexamethylenetetramine. Nitrogen include mixes be able to deteriorate yield smelling salts and extra conceivably unsafe unstable mixes. These resins require cautious dealing with measures. The cooked resin should be refrigerated awaiting the utilization of holding tanks and refrigerated trucks are mandatory. Indeed, even through refrigeration, the capacity duration of a phenolic gum is regularly 15 days.

11.9 Melamine Resins

This depends mainly on formaldehyde and 1,3,5-triazine-2, 4,6-triamine, which is as like urea-based formaldehyde polymers. Thus, mechanical utilization of this gum began in 1930's, the point at which the Swiss organization - Chemische Industrie Basel (CIBA) started the shifting from dicyandiamide to melamine. In olden days, the utilization was constrained due to its maximum cost. Presently, this can be delivered less expensive from urea-based polymers, and so the financial circumstance is enhanced.

11.9.1 Monomers

Melamine might be in part or completely supplanted with other reasonable amine containing mixes. Options in contrast to melamine incorporate ammelide, urea, substituted melamines, melem, melon, ammeline, guanamines, and thiourea. These mixes are likewise utilized as fire retardants. This substituted melamine incorporates mainly alkyl and aryl melamines. Delegate instances indicate that some alkyl-substituted melamines incorporate 1-methyl-3-propyl-5-butylmelamine, methylmelamine, ethyl melamine, trimethylmelamine, and dimethylmelamine. Foams and filaments show expanded flexibility, when a segment of melamine is supplanted by alternative melamine. In view of contemplations of expense and opportunity, normal melamine is commonly liked.

11.9.2 Other Modifiers

Ethylene urea, ethylene diamine, essential/auxiliary/tertiary amines, and melamine are the reasonable resin modifiers. Dicyandiamide will likewise fuse onto the resin. Therefore, the modifiers groupings in the reactive blend might change ordinarily between 0.05 and 5.00%. Every modifiers advance hydrolysis obstruction, lower formaldehyde discharges and polymer adaptability.

11.9.3 Synthesis

Like urea, this responds in the midst of formaldehyde in pitifully soluble fluid medium to frame methylol mixes. The further buildup continues under impartial and acidic conditions, along these lines producing methylene-based ether bonds. An unadulterated melamine-based resin was developed in a couple of existence at normal temperature range. In light of those undesired characteristics, these resins are mixed by urea resins.

11.9.4 Etherified Resins

These are set up via the response of melamine through formaldehyde underneath the states of pH about 6 and reflux temperature within the sight of a lot of butanol. Therefore, xylene cycles releases out the water framed by means of the buildup response utilizing azeotropic refining and quickened the etherification along these lines.

11.9.5 Properties

These resins are normally utilized for a few products which rely upon their ideal properties. Generally, these resins are solid, strong, and moderately shabby, yet are commonly shaded resins. However, melamine-based resins are water clear and are increasingly costly. They are commonly utilized for items, where the shading or example of the substrate is held with a straightforward melamine defensive covering or fastener. The emanation of formaldehyde in melamine resins is diminished by expanding the content of it. This is disclosed because of the solid holding between triazine carbons of melamine contrasted with urea carbons. These sulfonated resins show great dissolvability in water.

11.9.6 Applications and Uses

Resins based on melamine are broadly utilized as adhesives for wood, varnish, for enhancing the characteristics of paper and cellulosic materials, moldings, and as resins for ornamental covers. In contrast with urea formaldehyde, melamine-based gums have superior obstruction against warmth and dampness. So, etherified melamine resins are regularly utilized in blend by means of alkyd resins used for creation of brightening overlays. Change of materials via melamine is utilized toward bestow wrinkle opposition and shrinkage. However, the wet quality of paper is significantly enhanced by means of utilizing melamine resins as wet end added substances. Roof tiles were back covered by means of melamine resins so as to build them progressively inflexible and stickiness safe. Melamine resins are additionally utilized for the course of action of brightening paper covers. This function is because of their fantastic shading, chemical opposition, hardness, dissolvable and heat obstruction, and dampness opposition. Shaped articles like dinnerware are set up with a mix of urea and melamine formaldehyde resins. The resins are consolidated on account of the high costly melamine formaldehyde compound.

11.10 Furan Resins

The condensation reaction of furfuryl alcohol (FA) and obtained byproducts of that process are generally called furan resins. These resins are produced from vegetable cellulose and an inexhaustible asset. The major constituents present in the furans are also reviewed in detail. The modernization of furfural began in the year 1922 and the principal furan-based resins have come into sight.

11.10.1 Monomers

Furan, 5-methylfurfural, bis-2, 5-hydroxymethylfuran, furfuryl liquor, 5-hydroxymethylfurfural (HMF), 2-furfurylmethacrylate, and furfural are the most opted monomers for furan resins. The real favorable circumstances in this resins come along the way which resulted from vegetable based cellulose. Appropriate base of vegetable-based cellulose are oat bodies, corn cobs, sugar stick bagasse, biomass processing plant eluents, paper factory wastes, cottonseed structures, rice bodies, and some nourishment stuffs, i.e., saccharides and starch.

11.10.2 Furfural

As a whole, this type of aldehyde is a side-effect based on sugar stick bagasse, which generates resins by means of developing for a brilliant resistance to chemicals and low swelling. Furfural is produced using farming materials utilizing hydrolysis process. Generally, it is light yellow to golden hued straightforward fluid and turns dark browned colored. It is chiefly utilized in grease refinement, furfuryl liquor creation and pharmaceutical generation and used as central reagent to deliver materials, which contain monomers to deliver different furans in the midst of wide assortment substituents.

11.10.3 Furfuryl Alcohol

Furfuryl liquor is produced by means of reduction process with hydrogen. Generally, it is a dreary straightforward fluid and turns out to be light yellow or dark red or darker shading when uncovered noticeable all around. It varies with water and numerous natural solvents like liquor, ether, acetone, and so on, however not in hydrocarbon items.

11.10.4 Specialties, i.e., Polyimides Based on Furan

Polyimides dependent of poly (2-furanmethanol-formaldehyde) are capable to set up Diels-Alder response (DA) from separate furan through imides based on bismale. However, process continues inside tetrahydrofuran

(THF) otherwise in mass. THF intermediates and aromatize within sight of acidic anhydride. Therefore, these are dependent on top of this resin which displays great thermal dependability.

11.10.5 Special Additives as Reinforcing Materials

Aramid strands were utilized as fortifying substance for phenol gum and furan resin. Similar investigation of mechanical presentation of resources demonstrated that the resin is progressively reasonable as a grid contrasted with phenolic gum. The most significant mechanical furan resins depend on 2-furfuryl alcohol. So, the biggest makers of this resin are China and Dominican Republic. Around 30% of this resin usage is utilized to create alcohol, which is for the most part devoured via generation of furan-based resins.

11.10.6 Curing

Furan resins consist of inorganic and natural acids for curing. Instances of reasonable natural and inorganic acids consist of phosphoric, nitric, hydrochloric, tartaric, sulphuric, and maleic acids. Friedel-Crafts impetuses incorporate boron fluoride, aluminum trichloride, aluminum bromide, and zinc chloride. Resins along with enhanced imperviousness to flames are relieved by a blend of boric anhydride, trimethylborate, and p-toluene sulfonic acids. Both inorganic and natural acids with salts may likewise be utilized. Cyclic anhydrides like maleic anhydride were utilized as impetuses, which is trusted that co-polymers are framed among the anhydride and free hydroxylated species. Maleic acid advances the polymerization reactivity, which in turn specially diminish the outflow during curing process.

11.10.7 Recycling Properties

Extensive research led to bring swinging bunches into suitable polymers, for example, styrene by means of copolymerization with an appropriate co-monomer. Some authors talked about the impact of attractive nanoparticles that would improve the procedure capacity of reusing. The swinging furan is crosslinked by means of a bismaleimide to accomplish polymers through better execution. So as to reuse these crosslinked resources, heating trials through an overabundance of 2-methyl-furan have been performed to actuate the old age DA response of the system. Accordingly, the presentation of this resin component is a possible way of reusing crosslinked polymers through heat treatment by way of diene in abundance [40]. The DA response sandwiched with styrene-furfuryl methacrylate co-polymer tests and bismaleimide observed bright absorbance of maleimide bunch at 320 nm by utilizing through carbon-13 nuclear magnetic resonance (C-NMR) spectroscopy. This in turn explains the recycling properties of furan based resins.

11.10.8 Applications and Uses

Generally, these are utilized mostly in foundry business at the same time as sand fasteners used for throwing molds and centers. These are frequently utilized into mix through different resins, which are profoundly erosion safe. They have discovered to utilize in mortars and also in concretes with enhanced mechanical characteristics by fortifying through glass filaments.

11.10.8.1 Carbons

Permeable carbon like resins frames permeable carbon through pyrolysis process at a temperature of 450°C. However, glass-like structured carbon is distinguished as a brilliant carbon ancient rarity because of its attributes like hardness and shape strength. The non-graphitic arrangements of hexagonal sheets were found in the microstructure of glass like carbon. It has extraordinary properties like incredible hardness than other carbon materials and gases impermeability [41]. Glass carbon is set up via heat-treatment on thermosetting resins during idle climate. It has enthusiasm for the battery and semiconductor ventures.

11.10.8.2 Composite Carbon Fiber Materials

Carbon impregnation strands based on resulting pyrolysis process at 1,000°C enhance quality of carbon filaments. The utilization of an inactive impetus permits the use of a low thickness answer for the fiber by means of resulting polymerization at elevated temperatures. On the off chance that the forerunners were to polymerize fundamentally before its usage, the treating shower would be thick to the point which permits just a covering to be framed. In superior composite, carbonaceous material fortified with carbon fiber strengthened their strands. The parameters of permeable structure of this polymer are recorded in Table 11.5 [13].

Table 11.5 Parameters of porous structure of copolymers [13]. (With kind permission obtained from RightsLink, Elsevier License).

Polymer*	Surface area (m^2/g)	Average pore diameter (mm)	Specific pore volume (cm^3/g)
1.	70	160	0.12
2.	75	90	0.05
3.	130	178	0.1
4.	30	260	0.28
5.	11	nd	nd
6.	89	nd	nd
7.	68	nd	nd
8.	30	290	0.27
9.	20	250	0.14

*The polymers are the same described in Table 11.1, nd - no diameter.

11.10.8.3 Foundry Binders

Furans are expanded to some degree costlier than different covers; however, the likelihood of sand recovery is favorable. A standout among the most economically effective no-heat folios is the phenolic-urethane no-prepare fastener. This folio runs shape and centers with magnificent qualities that are created in an exceptionally profitable way. These have lesser VOC, low formaldehyde, free phenol level which generally generate a lesser amount of smell and burn acid center creation and castings. Conversely, restoring execution is greatly slower compared to curing of phenolic urethane fasteners. This can also altered to expand the reactivity by defining with suitable resins. All things considered, these changed fasteners don't give the fix speed required inside the foundries which require high profitability. An activator advances furfuryl liquor polymerization process was included. The restoring procedures of urea-based furan resins within sands were examined through infrared spectroscopy analysis.

11.10.8.4 Binders Based on Glass Fiber

In contrast, a suitable alternative is furan-based fasteners. Furan folios give a significant number of the benefits of phenolic fasteners while bringing about considerably diminished VOC discharges. Water as a remarkable part and emulsified furan resins can be utilized. Formaldehyde is certainly not a noteworthy restoring or disintegration side-effect and the furan resins structure exceptionally unbending thermosets. The furan related glass fiber restricting organizations are favorable because they permit the utilization of the same in having high sub-atomic loads or the expansion of different resources that would offer ascent to the arrangement of two-stage frameworks. An appropriate surfactant further to furan cover organizations is sodium dodecyl benzene sulfonate.

11.10.8.5 Oil Fields

Sandy wells and arrangement in oil-bearing were often hard to work on the grounds that the sand during development be ineffectively solidified and will in general stream onto the well along with the oil. However, generation of sand is a major issue, the sand results in disintegration, and untimely tiring out of siphoning gear. Normally, this is a disturbance towards expelling oil at a later stage as per the activity plan. This gum details are also utilized for *in situ* concoction sand solidification.

11.10.8.6 Substrates for Plant Growth

Customary mineral fleece plant development substrates depend mainly on sound network of mineral fleece in which the strands are commonly associated through a restored folio. There is also a need towards decreasing the phytotoxicity of synthetics utilized. This may resulted through the phenolic cover materials. On the off chance that a phenolic resin is utilized as fastener, a wetting operator is obliged to include request to bestow the hydrophobic mineral fleece framework with hydrophilic characteristics. In any case, the utilization of this resin permits the surrender utilization of particular wetting specialist. The main drawback is nearly of high cost. Hence, the conventional PF gum replaced by a furan resin is adequate towards accomplishing the ideal properties.

References

1. Clair, D.J.S., Star polymer adhesives, Google Patents, 1983.
2. Lee, S.-W., Park, J.-W., Kwon, Y.-E., Kim, S., Kim, H.-J., Kim, E.-A., Woo, H.-S., Swiderska, J., Optical properties and UV-curing behaviors of optically

clear semi-interpenetrated structured acrylic pressure sensitive adhesives. *Int. J. Adhes. Adhes.*, 38, 5–10, 2012, https://doi.org/10.1016/j.ijadhadh.2012.04.002.
3. Lee, S.-W., Park, J.-W., Park, C.-H., Kim, H.-J., Enhanced optical properties and thermal stability of optically clear adhesives. *Int. J. Adhes. Adhes.*, 50, 93–95, 2014, https://doi.org/10.1016/j.ijadhadh.2013.12.029.
4. Lee, S.-W., Park, J.-W., Park, C.-H., Kwon, Y.-E., Kim, H.-J., Kim, E.-A., Woo, H.-S., Schwartz, S., Rafailovich, M., Sokolov, J., Optical properties and UV-curing behaviors of optically clear PSA-TiO 2 nano-composites. *Int. J. Adhes. Adhes.*, 44, 200–208, 2013, https://doi.org/10.1016/j.ijadhadh.2013.03.001.
5. Lefebvre, K., Miscibility and influence on each other of a UV photo-initiated thermosetting polymer blend of radical and cationic systems, Mechanical (and Materials) Engineering – Dissertations, Theses, and Student Research, 50, 2013, http://digitalcommons.unl.edu/mechengdiss/50.
6. Baek, S.-S., Jang, S.-J., Lee, J.-H., Kho, D.-H., Lee, S.-H., Hwang, S.-H., Preparation of acrylic pressure sensitive adhesives for optical applications and their adhesion performance. *Polym.-Korea*, 38, 199–204, 2014, https://doi.org/10.7317/pk.2014.38.2.199.
7. Mangold, C., Wurum, F., Obermeier, B., Frey, H., Hetero-multifunctional poly (ethylene glycol) copolymers with multiple hydroxyl groups and a single terminal functionality. *Macromol. Rapid Commun.*, 31, 258–264, 2010, https://doi.org/10.1002/marc.200900472.
8. Bauer, M., Bauer, J., Kuhn, G., Kinetics and modelling of thermal polycyclotrimerization of aromatic dicyanates. *Acta Polym.*, 37, 715–719, 1986, https://doi.org/10.1002/actp.1986.010371112.
9. Otero, T.F. and Cantero, I., Conducting polymers as positive electrodes in rechargeable lithium-ion batteries. *J. Power Sources*, 81, 838–841, 1999, https://doi.org/10.1016/S0378-7753(98)00236-5.
10. Aksimentyeva, O., Konopelnyk, O., Bolesta, I., Karbovnyk, I., Poliovyi, D., Popov, A.I., Charge transport in electrically responsive polymer layers. *J. Phys.: Conf. Ser.*, 93, 012042, 2007, https://doi.org/10.1088/1742-6596/93/1/012042.
11. Sahin, E., Camurlu, P., Toppare, L., Both anodically and cathodically coloring electrochromic polymer based on dithieno [3,4-b,3′,4′-e]-[1,4]-dithiine. *Synth. Met.*, 156, 124–128, 2006, https://doi.org/10.1016/j.synthmet.2005.10.019.
12. Menger, F.M. and Tsuno, T., Synthesis of porous polystyrene with chemically active surfaces. *J. Am. Chem. Soc.*, 112, 6723–6724, 1990, https://doi.org/10.1021/ja00174a048.
13. de Santa Maria, L.C., Aguiar, M.R.M.P., Guimarães, P.I.C., Amorim, M.C.V., Costa, M.A.S., Almeida, R.S.M., Aguiar, A.P., Oliveira, A.J.B., Synthesis of crosslinked resin based on methacrylamide, styrene and divinylbenzene obtained from polymerization in aqueous suspension. *Eur. Polym. J.*, 39, 291–296, 2003, https://doi.org/10.1016/S0014-3057(02)00196-9.
14. Lin, T., Wang, L., Wang, X., Kaynak, A., Polymerising pyrrole on polyester textiles and controlling the conductivity through coating thickness. *Thin Solid Films*, 479, 77–82, 2005, https://doi.org/10.1016/j.tsf.2004.11.146.

15. Hanack, M., Intrinsic semiconducting materials on phthalocyanine basis. *Turk. J. Chem.*, 22, 13–22, 1998, https://dergipark.org.tr/tr/pub/tbtkchem/issue/11964/143158.
16. La Scala, J.J., Sands, J.M., Orlicki, J.A., Fatty acid-based monomers as styrene replacements for liquid molding resins. *Polym.*, 45, 7729–7737, 2004, https://doi.org/10.1016/j.polymer.2004.08.056.
17. Fors, B.P. and Hawker, C.J., Control of a living radical polymerization of methacrylates by light. *Angew. Chem. Int. Ed.*, 51, 8850–8853, 2012, https://doi.org/10.1002/anie.201203639.
18. Yamago, S. and Nakamura, Y., Recent progress in the use of photoirradiation in living radical polymerization. *Polym.*, 54, 981–994, 2013, https://doi.org/10.1016/j.polymer.2012.11.046.
19. Sasthav, M. and Cheung, H.M., Characterization and polymerization of middle-phase microemulsions in styrene/water systems. *Langmuir*, 7, 1378–1382, 1991, https://doi.org/10.1021/la00055a015.
20. Fyfe, C.A., Niu, J., Rettig, S.J., Burlinson, N.E., High-resolution carbon-13 and nitrogen-15 NMR investigations of the mechanism of the curing reactions of cyanate-based polymer resins in solution and the solid state. *Macromol.*, 25, 6289–6301, 1992, https://doi.org/10.1021/ma00049a028.
21. Li, Z. and Chau, Y., Synthesis of linear polyether polyol derivatives as new materials for bioconjugation. *Bioconjugate Chem.*, 20, 780–789, 2009, https://doi.org/10.1021/bc900036f.
22. Reddy, A.R. and Reddy, K.H., Heavy metal ion uptake properties of polystyrene-supported chelating polymer resins. *J. Chem. Sci.*, 115, 155–160, 2003, https://www.ias.ac.in/article/fulltext/jcsc/115/03/0155-0160.
23. Zang, D.H. and Lader, W., Polyisocyanate acrylate polymer adhesives, Google Patents, 1970, https://patents.google.com/patent/US3532652A/en.
24. Zheng, Y., Stevens, P.D., Gao, Y., Magnetic nanoparticles as an orthogonal support of polymer resins: Applications to solid-phase Suzuki cross-coupling reactions. *J. Org. Chem.*, 71, 537–542, 2006, https://doi.org/10.1021/jo051861z.
25. Berdahl, J.P., Johnson, C.S., Proia, A.D., Grinstaff, M.W., Kim, T., Comparison of sutures and dendritic polymer adhesives for corneal laceration repair in an in vivo chicken model. *Arch. Ophthalmol.*, 127, 442–447, 2009, https://doi.org/10.1001/archophthalmol.2008.582.
26. Carpi, F. and De Rossi, D., Colours from electroactive polymers: Electrochromic, electroluminescent and laser devices based on organic materials. *Opt. Laser Technol.*, 38, 292–305, 2006, https://doi.org/10.1016/j.optlastec.2005.06.019.
27. Hyun, K.S. and Spalding, M.A., Bulk density of solid polymer resins as a function of temperature and pressure. *Polym. Eng. Sci.*, 30, 571–576, 1990, https://doi.org/10.1002/pen.760301003.
28. Vadala, M.L., Thompson, M.S., Ashworth, M.A., Lin, Y., Vadala, T.P., Ragheb, R., Riffle, J.S., Heterobifunctional poly (ethylene oxide) oligomers

containing carboxylic acids. *Biomacromolecules*, 9, 1035–1043, 2008, https://doi.org/10.1021/bm701067d.
29. Thompson, M.S., Vadala, T.P., Vadala, M.L., Lin, Y., Riffle, J.S., Synthesis and applications of heterobifunctional poly (ethylene oxide) oligomers. *Polym.*, 49, 345–37, 3, 2008, https://doi.org/10.1016/j.polymer.2007.10.029.
30. Bobo, E., Development of a controllable polymer system using interpenetrating networks, Mechanical (and Materials) Engineering – Dissertations, Theses, and Student Research. 53, 2013, http://digitalcommons.unl.edu/mechengdiss/53.
31. Grenier-Loustalot, M.-F., Lartigau, C., Metras, F., Grenier, P., Mechanism of thermal polymerization of cyanate ester systems: Chromatographic and spectroscopic studies. *J. Polym. Sci. A: Polym. Chem.*, 34, 2955–2966, 1996, https://doi.org/10.1002/(SICI)1099-0518(199610)34:14<2955::AID-POLA13>3.0.CO;2-J.
32. Ohtsuki, A., Goto, A., Kaji, H., Visible-Light-Induced reversible complexation mediated living radical polymerization of methacrylates with organic catalysts. *Macromol.*, 46, 96–102, 2012, https://doi.org/10.1021/ma302244j.
33. Leporini, D., Zhu, X.X., Krause, M., Jeschke, G., Spiess, H.W., Probing porous polymer resins by high-field electron spin resonance spectroscopy. *Macromol.*, 35, 3977–3983, 2002, https://doi.org/10.1021/ma011861a.
34. Kim, J.-i., Kang, P.H., Nho, Y.C., Positive temperature coefficient behavior of polymer composites having a high melting temperature. *J. Appl. Polym. Sci.*, 92, 394–401, 2004, https://doi.org/10.1002/app.20064.
35. Zhu, X.X., Banana, K., Liu, H.Y., Krause, M., Yang, M., Cross-linked porous polymer resins with reverse micellar imprints: Factors affecting the porosity of the polymers. *Macromol.*, 32, 277–281, 1999, https://doi.org/10.1021/ma980599b.
36. Mythili, C.V., Malar Retna, A., Gopalakrishnan, S., Synthesis, mechanical, thermal and chemical properties of polyurethanes based on cardanol. *Bull. Mater. Sci.*, 27, 235–241, 2004, https://doi.org/10.1007/BF02708512.
37. van Dijken, J.W.V., Sjostrom, S., Wing, K., Development of gingivitis around different types of composite resin. *J. Clin. Periodontol.*, 14, 257–260, 1987, https://doi.org/10.1111/j.1600-051X.1987.tb01529.x.
38. Brochmann, J., Edwardson, C., Shmulsky, R., Influence of resin type and flake thickness on properties of OSB. *For. Prod. J.*, 54, 51–55, 2004, http://agris.fao.org/agris-search/search.do?recordID=US201300990828.
39. Edwards, W.L. and Schiavone, R.J., Polymer resins having reduced frictional properties, Google Patents, 2004.
40. Yoon, T.-J., Lee, W., Oh, Y.-S., Lee, J.-K., Magnetic nanoparticles as a catalyst vehicle for simple and easy recycling. *New J. Chem.*, 27, 227–229, 2003, https://doi.org/10.1039/B209391J.
41. Smigol, V., Svec, F., Frechet, J.M.J., Use of polymeric catalysts in the pore-size-specific functionalization of porous polymers. *Macromol.*, 26, 5615–5620, 1993, https://doi.org/10.1021/ma00073a013.

Index

Also of Interest

Check out these published related titles from Scrivener Publishing

Advances in Contact Angle, Wettability and Adhesion Volume 4
Edited by K.L. Mittal
Published 2019. ISBN 978-1-119-59254-9

Adhesives for Wood and Lignocellulosic Materials
By R.N. Kumar and A. Pizzi
Published 2019. ISBN 978-1-119-60543-0

Progress in Adhesion and Adhesives, Volume 4
Edited by K.L. Mittal
Published 2019. ISBN 978-1-119-62525-4

Advances in Contact Angle, Wettability and Adhesion Volume 3
Edited by K.L. Mittal
Published 2018. ISBN 978-1-119-45994-1

Laser Technology: Applications in Adhesion and Related Areas
K.L. Mittal and Wei-Sheng Lei
Published 2018. ISBN 978-1-119-18493-5

Progress in Adhesion and Adhesives, Volume 3
Edited by K.L. Mittal
Published 2018. ISBN 978-1-119-52629-2

Textile Finishing: Recent Developments and Future Trends
Edited by K.L. Mittal and Thomas Bahners
Published 2017. ISBN 978-1-119-42676-9

Progress in Adhesion and Adhesives, Volume 2
Edited by K.L. Mittal
Published 2017. ISBN 978-1-119-40638-9

Adhesion in Pharmaceutical, Biomedical and Dental Fields
Edited by K.L. Mittal and F.M. Etzler
Published 2017. ISBN: 978-1-119-32350-1

Interface/Interphase in Polymer Nanocomposites
Edited by Anil Netravali and K.L. Mittal
Published. 2016. ISBN 978-1-119-18491-1

Progress in Adhesion and Adhesives
Edited by K.L. Mittal
Published 2015. ISBN 978-1-119-16219-3

Advances in Contact Angle, Wettability and Adhesion Volume 2
Edited by K.L. Mittal
Published 2015. ISBN 978-1-119-11698-1

Particle Adhesion and Removal
Edited by K.L. Mittal and Ravi Jaiswal
Published 2014. ISBN 978-1118-83153-3

Laser Surface Modification and Adhesion
Edited by K.L. Mittal and Thomas Bahners
Published 2014. ISBN 978-1-118-83163-2

Adhesion in Microelectronics
Edited by K.L. Mittal and Tanweer Ahsan
Published 2014. ISBN 978-1-118-83133-5

Advances in Contact Angle, Wettability and Adhesion Volume 1
Edited by K.L. Mittal
Published 2013. ISBN 978-1-118-47292-7

Advances in Modeling and Design of Adhesively Bonded Systems
Edited by S. Kumar and K.L. Mittal
Published 2013. ISBN 978-1-118-68637-9

Atmospheric Pressure Plasma Treatment of Polymers
Edited by Michael Thomas and K.L. Mittal
Published 2013. ISBN 978-1-118-59621-0

www.scrivenerpublishing.com